Praise for *Intertidal*

'This is at once a work of detailed, shimmering natural history by an exceptional field naturalist and a profound meditation on what it means to be human – and more-than-human – in this damaged, astonishing world of ours ... Please read it!'

Robert Macfarlane, author of *Underland:*
A Deep Time Journey

'An impressive debut from an outstanding new author – a wonderfully thoughtful book'

Peter Frankopan, author of *The Silk Roads* and
The Earth Transformed

'*Intertidal* is the dazzling outcome of deep observation and magical writing. It reveals the wondrous diversity of liminal ocean life, and radiates into a fervent meditation on human consciousness itself'

Colin Thubron, author of *Shadow of the Silk Road*

'A wonderful and wonderfully talented new voice, gentle and poetic, subtle, watchful and observant, writing very much in the tradition of Robert Macfarlane and haunted by the ghosts of Barry Lopez and J.A. Baker. This is a startlingly brilliant and moving debut, written in pellucid prose as clear and inviting as a rock pool, as calming as the lapping of the sea on the shore, and as hopeful as a new dawn on the Coromandel coast'

William Dalrymple, author of *The Anarchy*

'A fine book'

Bruce Pascoe, author of *Dark Emu*

'This is a meditation on the world. Yuvan tells us where he's writing from: a pit, a forest, a room, a memory. Sometimes the entries are like a diary, sometimes a letter, sometimes an explication, other times a story ... often all of that, and more. The precision of his prose works like a hymn and a lament and a call to both attention and action'

Pádraig Ó Tuama, poet, presenter and
author of *Poetry Unbound*

'Truly inspirational—it should be read especially by any youngster who wishes to tread on a path less travelled, by all Chennai citizens and others who love to love nature'

Seetha Ananthasivan, founder trustee, Bhoomi College and
Prakriya Green Wisdom School

'This book will transform your worldview of not just Chennai but whichever urban area you inhabit. Through *Intertidal* you will discover new ways to see, hear, feel, sense—and care for—the natural wonders in your neighbourhood, and beyond'

Prerna Singh Bindra, author of *The Vanishing* and former
member, National Board for Wildlife

'Liltingly lyrical, rich in detail ... It is at the same time a deeply personal and authentic story'

Ravi Chellam, wildlife biologist and conservation scientist,
CEO, Metastring Foundation and coordinator,
Biodiversity Collaborative

'One of the most brilliant, gifted, observant and thorough nature writers of our times'

Rohan Chakravarty, cartoonist and creator of
Green Humour

'This is a book that deserves a special place in the shelves—and hearts—of all urban dwellers'

Shekar Dattatri, award-winning wildlife and conservation filmmaker

'*Intertidal* is at once a diary of healing, of action and of wonder. It is truly a transformative read'

Bathsheba Demuth, author of *Floating Coast* and associate professor of environmental history, Brown University

'An immersion in nature and a poignant account of the coastline, *Intertidal* sits on the intersection of our inner world and the biodiversity of the natural world. It will define the path of nature writing in the years to come'

Bahar Dutt, award-winning journalist and author of *Green Wars*

'The Coromandel Coast comes alive like never before in the moving meditations of a young naturalist highly sensitive to its fragile and wondrous ecology. His eloquent reflections pulsate with the dramatic interplay of human and more-than-human forces along the shore'

Meghaa Gupta, author of *Unearthed: An Environmental History of India*

'Yuvan's *Intertidal* is a DIY meditation manual, self-care resource and nature guidebook all rolled into one'

Nityanand Jayaraman, social and environmental activist

'*Intertidal* is a masterfully crafted invitation to pull close—in solidarity, friendship and studenthood—with our winged, pawed, petalled and shored sisters, to learn from their ancient, patient radicalisms, and to be transformed … a gift which will leave its every recipient nourished, breathing deeper and more alive'

'Ecological treatise, poetic invitation to nature-based meditation, quiet rage at the desecration of nature and displacement of ancient ways of life, and a subtly powerful plea to go beyond binaries and touch the intertidal zones each one of us inhabits'

'The voice of a new generation of Indian naturalists … ardent, informed, unafraid to confront the mess we've made—and are still making—of our landscapes'

'Yuvan paints a landscape of beauty and fragility. Every word is an expression of love for creation that also beseeches us to awaken to the ecological crisis that is at our doorstep'

'*Intertidal* is a treasure like no other. You cannot—must not— read it in a hurry. This book is to be read closely, mulled upon, meditated upon and savoured … for it shows us a way to be in this crazy, confusing world'

'Weaving stories of small creatures where the ocean meets land, Yuvan Aves takes us on a journey through truth, beauty and a relentless search for answers. This book offers hope, not only for individuals but for our species. A joy to read'

'Yuvan Aves's writing is emotive, immersive, almost pictorial in its beauty ... Do read *Intertidal* for hope, for joy, for wisdom and for courage and to fall in love with the rain, the oceans, the trees and the miracles that live and breathe all around us'

Dia Mirza, actor, producer, and Goodwill Ambassador, United Nations Environment Programme

'*Intertidal* is an extraordinary riptide of a memoir awash in place, environmental activism and biophilia—like a fresh sea breeze of environmentalism in India'

Aasheesh Pittie, author of *The Living Air* and editor, *Indian BIRDS*

'Yuvan's words always remind me that I'm an extension of nature and not a separate entity. His book *Intertidal* is thought-provoking and explores the connection between humans and the more-than-human. A must-read for everyone!'

Disha Ravi, climate activist

'This is a book of sensuous attention and tropical affection for the only thing that remains—this duniya, this nearest life'

Sumana Roy, author of *How I Became a Tree*

'A lyrical, meditative journey through coasts, estuaries and marshes, possibly the most human-abused ecosystems on the planet'

Bittu Sahgal, founder editor, *Sanctuary Asia*

'Searingly honest, deeply insightful, meditative and provocative all at the same time. *Intertidal* is an ode to a coastal lifescape that will stay with you for a long, long time. Diary writing can't get more lyrical than this'

Pankaj Sekhsaria, author of *Islands in Flux* and faculty at the Centre for Policy Studies, IIT Bombay

INTERTIDAL

INTERTIDAL

THE HIDDEN WORLD BETWEEN
LAND AND SEA

YUVAN AVES

ITHAKA

First published in the UK by Ithaka Press
An imprint of Bonnier Books UK

5th Floor, HYLO
103–105 Bunhill Row
London, EC1Y 8LZ

Owned by Bonnier Books
Sveavägen 56, Stockholm, Sweden

Hardback — 978-1-80418-981-8
Trade Paperback —978-1-80418-982-5
Ebook — 978-1-80418-983-2
Audio — 978-1-80418-984-9

A CIP catalogue of this book is available from the British Library.

Typeset by IDSUK (Data Connection) Ltd
Printed and bound by Clays Ltd, Elcograf S.p.A.

1 3 5 7 9 10 8 6 4 2

www.bonnierbooks.co.uk

For coasts, wetlands and all their life
For friends fighting to protect them

And for Rob and Rega

CONTENTS

Coast and Wetlands of Chennai and Surrounds

The map on the facing page depicts the formative features of Chennai's geography. They profoundly shape every aspect of its human and non-human life, history and future. The narrative of *Intertidal* flows along the coastline and waterways seen here.

1. Kattupalli Island
2. Ennore Creek
3. Marina Beach
4. Adyar Estuary
5. Kotturpuram Urban Forest
6. IIT Madras Campus
7. Guindy National Park
8. Great Banyan, Adyar
9. Urur Kuppam Beach (Elliot's Beach)
10. Kallukuttai Lake
11. Velachery
12. Pallikaranai Marsh
13. Chembarambakkam Lake
14. Thiruvanmiyur Beach
15. Neelankarai Beach
16. Pallavaram/Tirusulam Hills
17. Pallavaram Lake
18. Muttukadu Creek
19. Kovalam Beach
20. Kelambakkam Backwaters
21. Vellaputhur Lake

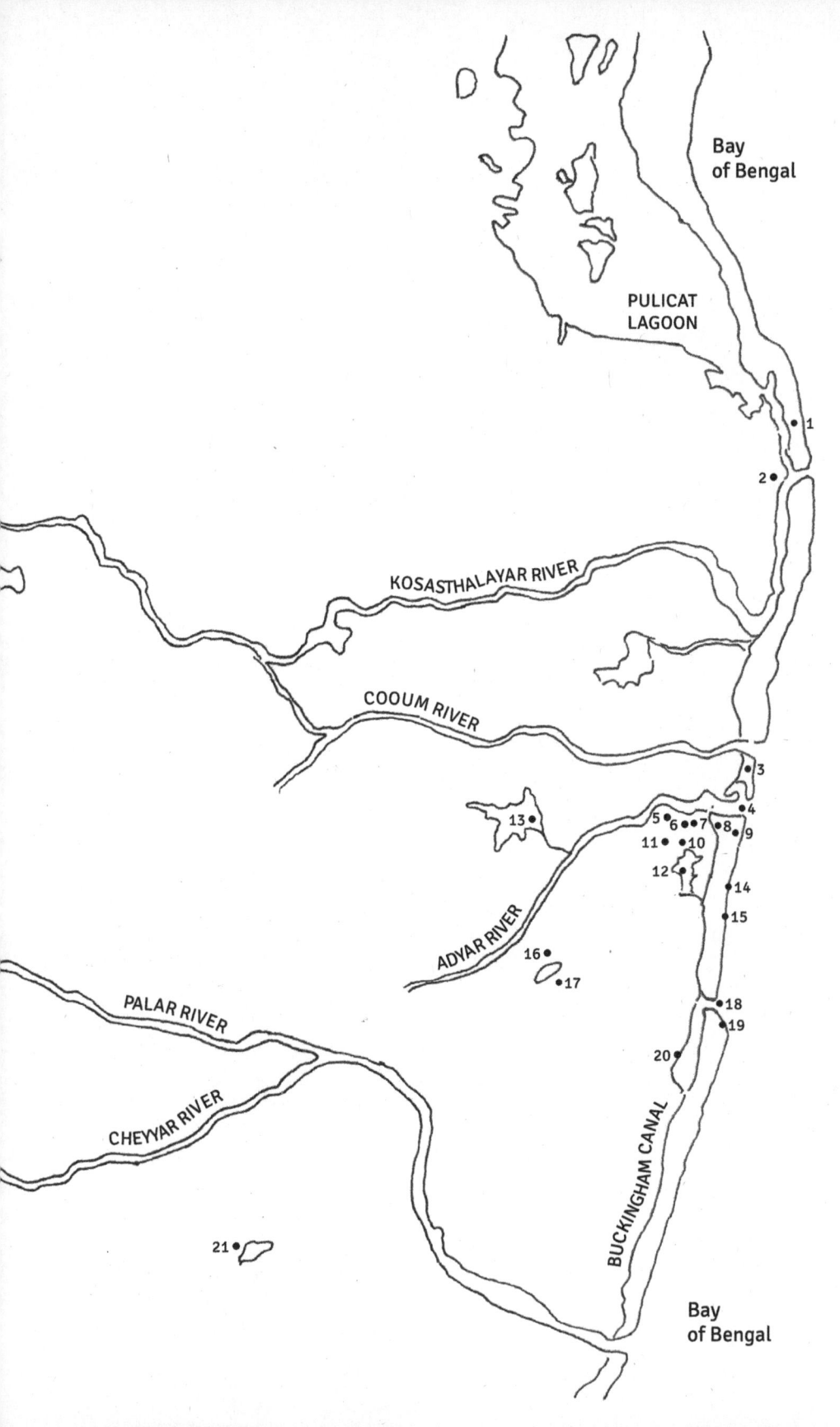

Bay
of Bengal

PULICAT
LAGOON

• 1

2 •

KOSASTHALAYAR RIVER

COOUM RIVER

• 3

• 4

13 •   5 • 6 • • 7 • 8
       11 • • 10   • 9

       12 •

                      • 14

                      • 15

ADYAR RIVER   16 •

PALAR RIVER   • 17

                    • 18

                    • 19

20 •

CHEYYAR RIVER

BUCKINGHAM CANAL

21 •

Bay
of Bengal

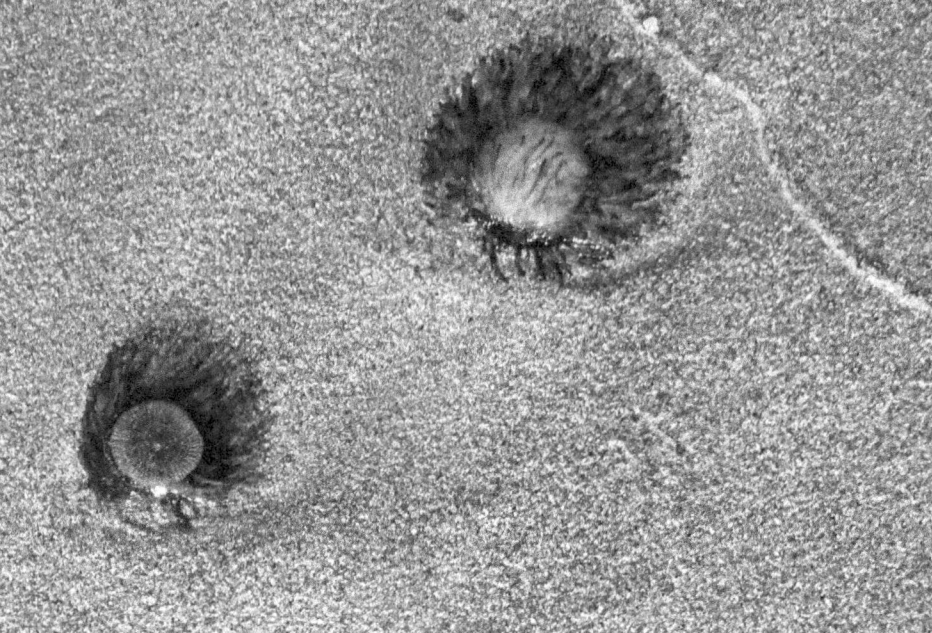

INTRODUCTION
BY ROBERT MACFARLANE

In the language of environmental science, an 'ecotone' is a transition zone between two ecological communities or habitats: where forest meets meadow, for instance, or where river marshes into pasture, or the shape-shifting realm where sea washes land. Ecotones are often vastly rich in terms of life. They are high in species abundance and diversity, and can also be what biologists call 'novelty centres' – that is to say, regions where speciation occurs at unusual rates.

Yuvan Aves's extraordinary book takes the intertidal ecotone at once as its site of study, its title, its literary form and its intellectual ethos. Like the strand line it explores, Yuvan's writing teems with life, complexity and novelty. It speciates. Its range of forms includes journal entries, glittering micro-poems, vivid *in situ* reportage, 'nature writing' and meditations. Its gauge of reference is thrillingly wide, connecting points familiar to British readers – John Locke, J.A. Baker, Nan Shepherd – with ancient Tamil poetry, the philosopher-speaker Jiddu Krishnamurti, the *Tao Te Ching*, Jungian psychoanalysts, physicist Giorgio Parisi's work on emergent properties and extended cognition, cutting-edge marine biological science or the political-theoretical work of Alexis Pauline Gumbs. Though *Intertidal* is utterly emplaced in the specificities of the Chennai shoreline, there is no more need for a reader to have first-hand acquaintance of that coast than there is to have visited Roanoke, Virginia, in order to love Annie

Dillard's *Pilgrim at Tinker Creek*, or the Shiant Islands to enjoy Adam Nicolson's gorgeous *Sea Room*.

Yuvan's writing is at once hospitable and welcoming – here is a person you'd love to walk beside, you think; whose eyes you feel lucky to look through – and beautifully disorientating in the strange affinities it discerns. I feel confident that the great Rachel Carson – another laureate of the intertidal, whose 'Sea Trilogy' lit up marine writing in the middle of the last century – would have cherished Yuvan's work for its combination of passion, knowledge and politics.

The web or rhizome is *Intertidal*'s model of relation, rather than the hierarchy. Everything here is in process, always-already in the midst of becoming-transforming-joining in some way or ways. Yuvan's vision of 'nature' is inclusive and egalitarian: it refuses to privilege size or supposed charisma, and is instead drawn to raising up the living world's invisibles, untouchables and inaudibles – humans and more-than-humans alike – such that they too become parts of '*palluyir*', a Tamil word meaning 'all of life'.

Take, for instance, Yuvan's four-page account here of the mass stranding of 'blue buttons' or 'sea swallows' on the Neelankarai Beach (the entry for 20 January 2022). It begins as a detective story: 'What has dumped them all on the shore?' There is a mystery to solve. Yuvan walks, inspects, reflects, scrutinises, hypothesises. Out of this one puzzle, he spins a remarkable meditation on how ocean life exists communally and mutually as well as competitively. 'A blue button', writes Yuvan:

> is not a single creature. Each seemingly dime-sized organism is a free-floating carnivorous creature-village made of polyps. Polyps themselves are small, jellyfish-like beings. These too are a puzzle between singular and plural, queer creatures, beyond the scope of our pronouns.

The 'carnivorous creature-village' of each blue button is then located within the broader 'surface community' of the ocean: its purple storm snails, goose barnacles, orange sea nettles and trevallies, all the way out (not up) to the fisherfolk, whose knowledge of the seascape is profound and precise. Where does one organism stop and the next begin, one wonders while reading this? New ways of seeing disclose themselves to the reader; emergent properties of these fast-shifting, quick-weaving sentences and ideas. The writing lives evocatively in the time and place of the stranding ('Stubby waves roll in slowly, one by one and far apart'), while also somehow keeping in view such widely distributed hyperobjects as garbage patches, earthquakes and ocean gyres.

Yuvan's expertise is indisputable; his humility unfeigned and thoroughgoing: 'It is overwhelming, the stories the ocean keeps bringing. Day by day it becomes vaster, more wonderful and mysterious.' Eventually, the question which set all of this in motion goes unsolved – but this is no failure, for the impulse to answer it has woven wonders together. Here, as across the book, the 'intertidal' serves as a portal or threshold, which allows movement back and forth between the material and the metaphorical-metaphysical, which are of course eventually understood to be inextricable: 'inner and outer merge, and one simultaneously walks on all shores', as Yuvan puts it. Here, too, we see in action an ardent quest to understand more, without the faintest presumption that one might ever understand all. As the Scottish writer Nan Shepherd concisely phrased it in *The Living Mountain*, a book dear both to Yuvan and to me, 'The thing to be known grows with the knowing'.

I am fortunate enough to have walked the Chennai coastline with Yuvan. It was an unforgettable experience. He is flat out and hands down the best all-round naturalist with whom I have ever spent time in the field. Fungi, protists, plants, snakes, birds, fish, insects, molluscs, lichens, humans: his knowledge spans the many kingdoms (or as biologist Giuliana Furci prefers to call them, 'kin-doms') and phyla of life. Walking with him and listening to him relate – in two senses of the verb – what we were seeing and moving through, the world sprang into new brightness and subtlety; became what the philosopher A.N. Whitehead called 'intervolved'.

That brightness sings in his prose, too. *Intertidal* is a deeply thoughtful work; it also *glitters* at the level of the image. I could quote dozens of examples, but here are a few of my favourite field-note gleanings:

'July rains in Chennai have been coming like golden jackals – loud and mostly at night.'

'The winter sun was two fist-spans over the ocean and pleasantly warm.'

'I found a single whale vertebra on the beach, thick as a palm trunk, heavy as a shot-put ball.'

'The paddy fields are booming. The frog call drums on your skull.'

Sometimes, Yuvan's writing is mischievous or provocative, inhabiting unusual temporalities or kinships, bringing you to think in strange ways:

'A dune lives a slow life, one which is difficult to observe within a single human lifetime.'

'… any religion can be turned into bark, leaves, heartwood and sapwood.'

'Fruit-flies can feel lonely and lose sleep over not being with their friends.'

'I have lunch sitting with the bee sisters at noon today, between back-to-back classes.'

Always, always it is alert to the complexities, difficulties and possibilities of how humans and other-than-humans might live together on this Earth.

I want to end this heartfelt praise-song of a foreword by turning to the politics of this book. 'There is a kind of attention', Yuvan writes early on, 'that does not separate itself from action.' If *Intertidal* has a doctrine, this is it. Aesthetics unfastened to ethics is little but luxury; identification alone will change nothing. Across its course, the book seeks to draw 'operative democratic lessons', as Yuvan puts it, from the living world, and to make injustice audible even where power seeks to shroud it in silence.

Today's India is a dangerous place to be an environmental activist. Those who speak out against the double-monolith of the BJP and the development agenda – Big Modi and Big Money – place themselves in jeopardy; get marked as 'anti-India'. In 2021, India was ranked the third most dangerous country in the world to be an environmental protester, after Mexico and Colombia. In 2023 it was ranked 161 out of 180 countries in the World Press Freedom Index. Yuvan has already faced down a number of hostile manoeuvres by state authorities seeking to undermine or restrict his work as an activist.

I have seen Yuvan in political action first-hand. One day during my time with him in Chennai, we travelled to Odiyur lagoon, a water body threatened by a road-widening project. We painstakingly gathered empirical evidence (bird counts; presence of seagrass; video footage of other possible, less destructive routes for the road), then met with a fast-talking, high-octane environmental lawyer called Yogeshwaran Amarneethi. Yogesh submitted the package of evidence to the courts the next morning. Within two days the judge had ruled that the highway project must not impinge on the lagoon. 'The birds swayed the bench!', Yuvan messaged me in delight. Attention had become action.

Yuvan's work – the work of his life, of this book – is that of both enlivement and protection. The world that he perceives, and shares with others through his teaching, activism and writing, is a world of far more life and lives – distributed across species boundaries and clades, and up and down scales – than power typically allows. He is, we could say, a counter-mapper, whose counter-maps render visible the inextricable intervolvement of all beings.

Here, then, is a book which walks us through an intertidal world into which both feet and thought sink, and where binaries are dissolved. It proves a hundred times over the definition of 'love' as meaning 'to live in intimate observation'.

—— Robert Macfarlane

PROLOGUE

High-tide sea. Early April wind blows feebly from the southeast. Crows sit airily along the sterns of boats facing the sea as if there is nothing for them to do that day. Waves curl south to north, stooping into question marks then sloshing on the intertidal from side to side. Ocean is made mercury-slick and silvery-opaque by the low glow of cloud-hidden sun. Two humming blue bullets shoot into the sea, one after another, and curve up into sky. A pair of carpenter bees. I've brought a group of teachers on a shore walk to Urur Kuppam Beach,[1] right after finishing the final draft of this book. We sit at the boat dockyard around shells, shards and other oddities. Each of us sketches a mini field guide of our findings—four to a page—and shares it with one another.

Some fishermen are returning from the sea, their yellow and green sardine nets rolled up in a heap in the middle of their boats. I am showing my colleagues ways of making the coast a living, learning space for children. On the sand are an unusually high number of egg sacs of marine sea snails. We peer into the chambers of a spindle snail's egg sac, which is a long cord with interlocking, clam-like purses all around it. Every purse has a few dozen tiny spindle snails, each snail a clove's length. Five brown-headed gulls in full breeding plumage squawk as they fly over the fishing hamlet and into the ocean, possibly beginning their return journey far north, having spent the winter with us. When I mention this, one of my fellow teachers waves the birds farewell.

[1] Also known by its colonial name Elliot's Beach.

I've been coming to Urur Kuppam Beach since I was a child, like most residents of Chennai. On Sundays the beach is one of the most crowded places in the city. On less busy days and nights its other denizens are more active—ghost crabs, wedge clams, olive ridley sea turtles, sand plovers, olive sea snails, mole crabs and more. In a way, as a matter of practice, it is home for me—after taking people on numerous shore walks, working with the fisher community, organizing rallies and campaigns on the beach's promenade and also just coming here alone to spend time with the sea. The campus of the school I grew up in was near this beach and within reach of the sea wind.

The School (and that is its name) was once here, in the forested lands of the Theosophical Society. The most charismatic entity within campus to me was a large banyan tree near the junior quadrangle and kindergarten which was its own village of creatures. There was also a pond in school, which was out of bounds for students, but where you could see flapshell turtles and elusive birds like the orange-headed ground thrush, forest wagtail, chestnut-winged cuckoo and the Indian pitta. I'd often get into trouble with teachers for venturing out there. One day after school I sat by the pond watching a shikra leaping back and forth across the waterbody between two large rain trees catching dragonflies that were hovering over it, and forgot about home. They had a search party roaming across school for me.

I learnt to count, add and subtract with seeds from the coralwood, gulmohar and rosary pea vine. School was my safe place, while family felt like a war zone. I took root in the philosophical foundation which J. Krishnamurti had laid: 'Question every answer', 'To be is to be related', and 'Be a light unto yourself'. But the deeper philosophy was the space itself—a forest within a city.

My mother had nurtured my eagerness for nature since I was born. She had observed me as a toddler going after insects or keenly observing leaves falling from the rain tree near home and would take me to wild places and get books and materials on other species for me to learn. Her practice of parenting during my early childhood was 'following the child'. She often tells me and other people that I had an inborn connection with and curiosity for the more-than-human. But I would later find, as a nature educator having worked with thousands of children, every child from under the age of seven to ten years old has this interest in nature and an intrinsic wildness along with a voracious capacity to wonder. This is then erased or suppressed by a number of factors on the path to adulthood.

Despite my mother's care, family was always a very difficult place for me through childhood. My biological father did not take much interest in caring for us. He'd leave home for weeks or months, often with other women. This would cause frequent fights and constant unrest at home, and my biological father would hit my mother for questioning his ways. Then for months together we'd have no money for food or basic supplies and we'd depend on the generosity of my maternal uncle. In the next couple of paragraphs I'll be recounting the darkest period so far in my life. There are some descriptions of violence. But it is crucial to narrate these experiences I've had, the wounds I sustained and later was able to recast, in order to set things up for this book.

When I was eleven years old my mother initiated separation and divorce with my biological father and moved in with a new man—at the time a self-proclaimed naturopath and astrologer—who became my stepfather. He hated me for a number of reasons he'd constantly list, some of them being: I resembled my biological father, I was born under a star shared by all his enemies, I had the horoscope of a mentally retarded person and so on. He used me largely for domestic work—sweeping and mopping the house every day, washing the vessels, washing everybody's clothes, washing the bathrooms, doing the cooking along with buying cigarettes for him every morning before going to school. Then he'd try to find some trivial excuse to keep me from going to school. He'd beat me up at least five days a week on the smallest of whims. I'd have torn lips and wounds on other parts of my body most days of the month. He had the habit of continuing to beat me for hours through the night, and blood from my mouth and face would splatter on the house walls, making red stains which would begin to smell.

On weekends I would have to get up early and scrape my blood off the walls with a knife. Because if a visitor spotted it, I'd be in greater trouble. I was barely sent to school during these years but school was kind to me, sensing unrest at home and giving me extra time, space and care when I needed it. I couldn't be specific to my teachers about what was going on because that'd mean more trouble. On most days I slept for three or four hours due to overwork and from fear, because he'd often let me fall asleep and then kick me awake to start beating me up. Once he jammed a gate on my face and my left eye filled with fluid; I had trouble seeing clearly for months. Another time when I was pinned to the floor and kicked repeatedly on the side of the head, my eardrum tore and fluid began to leak. I felt hazy and was taken to the doctor, who said I needed surgery immediately. But stepfather didn't allow it. On one occasion I was made to tear my biology textbook— my favourite subject—page by page and put it in the fire over the stove.

We lived in Madipakkam, a place built entirely upon marshland seething with life. The violence inside the house and the urbanizing of the marsh outside echoed each other. However, the vestiges of what was left of the marsh were my refuge. Whenever I got out of the house, to walk to a shop or for any other reason, the marsh helped displace pain. Purple swamphen, Eurasian moorhen, painted stork, clamorous reed warbler, chestnut munia, yellow bittern, scarlet skimmer dragonflies, emerald spreadwing damselflies and pond-skaters were daily sights in the tall typha reed beds. In the winter I looked forward to the yellow wagtails, paradise flycatchers, skyfuls of bee-eaters in the evening glow and the many flocks of waders I couldn't yet identify.

As I walked from home to marsh, I crossed an intertidal space within and the world would completely change for me for a brief time. Occasionally, a peregrine falcon would appear in the sky and lift me up. Or a grey mongoose would cross my path. In the monsoon I was excited to see frogs and snakes. But my stepfather was a highly superstitious man with his own history of trauma and violence (as I describe later in Section 4: Detritivore Meditation). If he saw a snake he'd get somebody to kill it and then come home and bathe in water mixed with turmeric, rock salt and lemon. When I was fourteen, a wolf snake came into the house on a rainy day and he asked me to kill it. I took it and left it safely outside in the grass, came in and emphatically said that he could beat me all he wanted but it wouldn't make me kill a snake. I got beaten all through that night until dawn and thrashed with every piece of furniture in the house.

This was my life for close to five years. My mother would cry, pained by the violence I was going through, but she believed that her entire self-worth and existence came from the man of the house (even though she was a lawyer and the breadwinner, and this man would hardly leave the house to do any work or to meet the world) and that she was nothing without him. Today, she is a leading example for me of complete transformation, and somebody who has helped hundreds of women legally to be freed from their misery, especially from the tyrannical patriarchy in their families, similar to what we went through. My mother and I collaborate on many projects together now. But it was not so at that time.

A few days after I turned sixteen I ran away from home to escape the violence. In the evening of 11 September 2011 my stepfather decided that I'd no longer go to school and be permanently kept at home for domestic work. I grabbed some coin change for the bus and set off. Anywhere but here was the thought in my head—a forest, an ashram, some hill, anywhere else. It was when the bus was veering towards Besant Nagar that I actually started thinking about what it was I was trying to do. I got off the bus and stood on the road turning into my school a little before the Adyar River. At nine in the night I knocked on the door of G. Gautama, the school's director at the time. He used to live on campus. He welcomed me in, asked what the matter was and listened carefully to my story and my decision. The next day he and the school administration spoke to my mother, who, though shaken by my leaving, was honest with the school about my suffering. Gautama said that the violence stopped for me that day and I would no longer be subject to it. He offered to take me in as an independent student at Pathashaala, a new residential school under the Krishnamurti Foundation that he was heading.[2]

Pathashaala is a hundred-acre residential school on the banks of the Palar River. The five years I spent there was a time of pupation and self-forging. It was a period of healing, but also of recasting the meaning of my own suffering and changing its architecture within me. Psychological impacts as such don't vanish, I've learnt, but their form and meaning in the mind can be changed into something fertile after a lot of hard work on oneself. On this vast campus and its surroundings, with its land largely left open to sky, I began to obsessively observe other species every day—birds, butterflies, snakes, trees, fungi, frogs, all life. To the south and west of the campus was the Palar River snaking towards the sea; to the east was the Vazhuvathur eri (lake) and the Salur reserve forest, which had a small hill I walked up often and where a pair of great horned owls lived for many years. To the north were nearly boundless paddy fields where I'd see five species of harriers quarter over during winter. Immediately south

[2] As an independent student I did not have to attend scheduled classes. I learnt by myself.

of the campus was the Vallipuram eri along the bund of which through the year I'd see eighty species of butterflies. I'd have conversations with everybody from children to teachers, farmers and administration staff, and I learnt to listen and meet different realities.

Seeing my keenness for the more-than-human and its entanglement with the human, the headmaster of Pathashaala, K. Ramesh, also a haiku poet and friend, gave me three books of nature writing which turned out to be profoundly formative for me. M. Krishnan's *Of Birds and Birdsong*, Robert Macfarlane's *Old Ways* and Annie Dillard's *Pilgrim at Tinker Creek*. These authors became lodestars, distant mentors. Krishnan taught me to observe and honour nature present around me. His writing evoked the richness of Indian landscapes. Dillard's raw authenticity with words helped me find my own. Macfarlane accompanied me on long walks and widened spatially and temporally my vision of landscape. He helped me grow into the intertidal spaces between the lived ecological terrain and my own imagination and heart. Seven years later I would begin a deep and growing friendship with him after reading his book *The Lost Words* and reaching out to him.

These writers (and later others) I read side by side while practising daily observation periods. I'd also ardently play my recorder, a woodwind instrument, teach it to a few students and compose music, some of it using the thematic phrases of common songbirds around me. My commitment to become a naturalist grew, which was for me not a profession, but a constellation of values, a worldview, a way of living consciously emplaced in a multi-being world—in 'a world of many worlds', as written in the Fourth Declaration of the Lacandon Jungle. I had been vocal about my unconventional aspirations when I was a child, but was forced to keep them suppressed for the ridicule they received. When I was among a group of lawyers in the chambers of the Madras High Court at the time my parents were getting divorced, I was asked what I wanted to become when I grew up. I unhesitatingly said ecologist, environmentalist, activist. The whole group of men burst out laughing. This was no longer the case. Pathashaala generously nurtured my innermost calling.

I'd go on long walks by the lakes, forests, hills and paddy fields in the Palar's basin. For a long period I suffered acutely from the years of violence I had been subjected to. Nightmares would make me fearful of going to sleep. I'd wake up in the night with intense fear, rage or in panic. Paying deep attention to nature as a daily practice helped me get out of my head and let light into my wounded mind. Wild places and other beings were an anchor through storms triggered by past violence; they provided resilience during the breakdown of my inner climate. As did the practices of reading and writing daily at the intertidal space of inner and outer observation. This book in large part is a sharing of this practice of deep seeing. This has been foundational in recasting my suffering into life-driving energy.

I began my work as a nature educator in the Vallipuram government school outside Pathashaala. On my walks some of the schoolchildren and village children would join me. At first they found it weird seeing somebody roaming around most of the time with a camera or binoculars in hand (both of which I'd borrowed from the teachers at Pathashaala). Eventually they got interested in the birds and butterflies and joined me. They learnt the names and shared stories and independent observations they'd make near their homes. Through Pathashaala's outreach department I put together a programme for three schools in the surrounding villages—Vallipuram, P.V. Kalathur and Aanoor—in which I made the local landscape the learning context. The programme ran quite well and many positive outcomes were observed in the children.

During this time I was pursuing my A-levels independently, studying by myself, getting some guidance from teachers at The

School and writing my exams at a different centre, as well as taking classes for children at Pathashaala. After the A-levels, when I was thinking about leaving for college, Gautama suggested that I could stay for a few more years and continue my work there while also pursuing the different things that interested me and studying for a college degree through distance education. I started with zoology at the Indira Gandhi National Open University but finished with a bachelor's degree in physics, having gotten interested midway in the intersections between quantum mechanics and consciousness and even ecology from reading people like Fritjof Capra, Gary Zukav and Karen Barad. My course had no such parallels. In 2016 I returned to Chennai to complete my degree and also to see how to take life forward from the various offbeat and precarious decisions I had taken in the past.

Human history tells us stories of 'battles, emperors, and land grabs'.[3] But it is our relationship with other beings that changes us and the planet. Any history of place or human community is equally, if not better, explained through non-human agencies. So is the case with Chennai. Cities emerge around hydrologies. Sometimes they remember their bedrock and live for long. And at other times they forget their watery genesis and collapse under their own weight. Three rivers flow through Chennai: the Adyar which is part of the Palar River's basin, the Kotralayar, the largest of the city's rivers, and the Cooum, which is really a distributary of the Kotralayar. (Cooum has become a pejorative for filth and stench.) In 1980, 80 per cent of Chennai's landscape was made up of wetlands and only 20

[3] https://www.waterstones.com/blog/keggie-carew-on-the-image-that-inspired-beastly#:~:text=History%20is%20all%20too%20human,have%20got%20far%20without%20them.

per cent of the city was built-up area. In 1991 the built-up area increased to almost the same as that of the wetland area. In 2010 only 15 per cent of Chennai's landscape constituted wetlands, the rest was built-up area. This figure has dwindled considerably since, coinciding with intense floods, cyclones during monsoons and acute water scarcity in summer. On 19 June 2019 the Tamil Nadu government declared 'Day Zero' when Chennai officially ran out of water due to multiple failed monsoons and had to buy water from other states.

The city that is Chennai was previously called Madras, and the etymology of this word is thought to have come from Madrasapattinam, a fishing village in this region before the city came into being. Some sources say it came from the name of a fisherman, Madrasan, who was the village headman. The city began to be established in 1639 as a trading settlement for the British East India Company on the eastern coast of the Indian peninsula. This later became the capital of the Madras Province under the British. Many of the place names around Madras that is now Chennai have the suffixes -kuppam and -pakkam, denoting presence by the sea or seaside. Place names with the suffix -eri denote the presence of a lake and people's dependence on it. All stories from Chennai must contain the rivers and the sea and their many forms of agency, and if they do not, there is surely something missing.

The British built Fort St. George, now used as the secretariat for the state of Tamil Nadu, and called its confines White Town and the places surrounding it Black Town, where local people lived and where they brought in large numbers of working-class people to labour in industries. Industrial expansion continued in independent India into the wetlands and backwaters of the Kotralayar River. The abuse of this landscape persisted. It is Chennai's largest sacrifice zone (places, people and non-people who are invisibilized/decimated in order for a city to exist). In this book I describe many journeys into this space with different people to meet its realities. And the implications of the desecration of the intertidal spaces of this region for the rest of the city, for its people and more than people.

According to a Council on Energy, Environment and Water report, Chennai is the second-most climate-vulnerable place in India. It is the most vulnerable specifically with regard to flooding, strong weather events and sea-level rise. In September 2022, the Greater Chennai Corporation released a Climate Action Plan for Chennai, which showed unsettling data for the city's climatic future. This included a 7-centimetre sea-level rise in five years, the sea ingressing 100 metres into the coast, the freshwater marshlands of Pallikaranai becoming brackish, 30 per cent of the city being inundated during monsoons in the next five years, heat stress and water stress significantly increasing in summer and so on. And as in any climate scenario, here too the entities causing the crisis are least affected by it, and those who have little to do with it—the historically marginalized, the non-human and the future generations—are the ones devastated. The 'hope in the dark', as Rebecca Solnit put it, for this region I think are the many youth movements, fresh political thinking and new imaginaries emerging to recast the city's suffering into something better.

———

Returning to Chennai in January 2016 I began compiling my writings and published my first book of essays, *A Naturalist's Journal*. I got in touch with Vijay Kumar, the secretary of the Madras Naturalists' Society (MNS), and became a regular part of their activities. I'd visit Anna Centenary Library and places like Kotturpuram Urban Forest, Perumbakkam marshland and Urur Kuppam Beach frequently—keeping myself sane and posting every day about nature and other species on Instagram and Facebook, sometimes twice a day. I joined Abacus Montessori School, introduced by my music teacher

S. Balakrishnan, and first began teaching recorder there but quickly joined their farm programme and started working as a nature educator. In a couple of years, I expanded it with the help of my colleagues into a farm, environment and society programme.

In 2018 I wanted to be an active part of the climate movement and I joined Fridays for Future India through which we took on several campaigns to save wild places being rampantly cleared across the country, especially with the onset of COVID-19 when people were most occupied and distracted with the pandemic. I sought out Nityanand Jayaraman, among the city's foremost environmental activists, whom I had heard speak during my school days about the Kodaikanal mercury poisoning by Hindustan Unilever. With his guidance and with friends from the Vettiver Collective and similar groups we formed the Chennai Climate Action Group and organized Chennai's first climate rally on the promenade road of Urur Kuppam Beach in 2019 and every year since. We work to unite the rights of fishers, labourers, minorities and historically marginalized communities with the ecological issues in Chennai, which are all fundamentally connected.[4]

One of the largest campaigns we worked on at the Chennai Climate Action Group is the Save Pulicat campaign—for the Pulicat Lagoon threatened with erasure by the 6,111-acre megaport proposed within the eco-sensitive zone of the Pulicat bird sanctuary. Pulicat is the second-largest brackish lagoon in India and the Ennore–Pulicat wetlands are Chennai's most important flood catchment and wetland system, into which flows the Kotralayar River. We undertook a number of field surveys and worked with local fisherfolk and people to strengthen the

4 'The economic system that's driving ecological breakdown is the very same system that also exploits workers, and women, and people of colour. These struggles are connected.' https://twitter.com/jasonhickel/status/1653745641363759105

movement. At the end of 2020—with the pandemic waxing and waning like spring and neap tides—my naturalist friends Vikas, Aswathi, Rohith, Nanditha and Anooja and I, through the MNS, began a survey of the coastal biodiversity hotspots of north Tamil Nadu—Pulicat, Adyar Estuary, Kovalam Creek, Odiyur Lagoon and Kaliveli Estuary. These habitats are crucial climate sanctuaries for Chennai but are all variously threatened. We documented the biodiversity, local fisherfolk knowledge and practices and threats to the ecosystem in each of these spaces. We built a scientific database through publications, nature-educational material and—importantly— citizen science which helped us later in the battles, ranging from public campaigns to court cases, to protect these areas from destructive discriminatory development.

We learnt greatly and collaborated on many fronts with fisher communities in these struggles, my greatest teacher among them Palayam Anna from Urur Kuppam, whose teachings about wind, weather and fish transformed my experience of the beach and the sea. I travelled to other parts of the Tamil Nadu coast to understand their local ecologies, complexities and conflicts, and to see if we could engage with or give solidarity to their campaigns. Some of these journeys are in this book as well.

In July 2021, through the MNS, I began running an internship for colleges in Chennai, with the support of Professor Kalpana Jayaraman, to train young life-science students as nature educators. The dream of this programme is to create a citywide network of young naturalists and facilitators around Chennai's landscape who could actively engage the public in their localities and possibly tilt the city's culture towards that of eco-literacy and belonging. They would help citizens become enmeshed in caring for and developing an interest in this unique landscape and bioregion. There are references in the chapters to this programme. After the first batch graduated, several interns expressed a wish to pursue nature-education activities in the city. They devoted their time to a number of projects even after the course was over. Their energy and also perhaps the city's dire need for more nature educators pushed me to start the Palluyir Trust for Nature Education and Research in December 2021. Through it, among many other activities, we started a Youth Climate Internship for youth from three climate-vulnerable communities—Urur Kuppam, Ramapuram and Kakkan Colony—to support and train them as nature educators, changemakers and justice facilitators in Chennai. Many of our sessions happen on Urur Kuppam Beach.

———

Intertidal is the part of the shoreline that appears during low tide and is hidden during high tide. In some places it is thin, in others it is vast. It can also be a metaphorical and metaphysical space. This is the theme and rhythm of this book, which is a diary of deep observation of the living world in my coastal city over two years, from 2020 to 2022, and through three monsoons. It comprises observations of coast, wetland, climate and self, and accounts of taking action to stand up for these life-sustaining spaces with my community of friends, co-workers and many, many teachers.

There is a kind of attention that doesn't separate itself from action. The text travels through the many campaigns and rallies we organized and the continued fight for coastal landscapes. This observation-action is my daily sacrality, my animistic practice. Every chapter is called a meditation with a broad theme aligned with the climatic season, and as in any meditation paying attention is the central action. There is a guided, nature-based, sensorial meditation in each chapter which appears like a tide somewhere during its length. Intertidals exist between beach and ocean, land and mind, other and self, and so on. Intertidals represent the habitats of the ambivalent and the unknown/unknowable. While observing, listening and perceiving, we are automatically in an intertidal space as one weaves into another and interlocks perceptions.

The book is set in Urur Kuppam and other beaches, the Velachery marshland where my home is, and in the many still-wild spaces of Chennai amidst its utter urbanity, where everything we do, think and choose is impacted by the fact that we are right by the ocean. At the same time it intersects—intertides—with the many events and realities within my mind which I have learnt to observe in continuum with the land and ocean. Wayfaring the mind is as vast, wondrous and challenging as wayfaring the coast and the ocean. For many of us these embarkings are the same. Inner and outer merge, and one simultaneously walks on all shores.

3 April–15 May 2023
Velachery–Pune–Velachery

Section 1

OCEAN MEDITATION

Urur Kuppam Beach

I sit facing the sea at Urur Kuppam and watch the dark waves. The spilling froth is slightly luminescent, and hundreds of large ghost crabs roam the intertidal. Urur Kuppam Beach is crowded in the evenings. I get a whiff of hot cotton candy. In the sea, trawler lights are closer to the shore than they should be. I found a beached guitarfish (*panaigyan* in Tamil) earlier, just under the high tide line, its sunken eyes waiting patiently for a wave to surge forward and take it back in. I carried it and left it in the water. It stayed still for a while. When the fourth wave lapped over it, it darted into the ocean with side flicks of its tail. Then I sat with eyes closed and let wave sounds permeate my mind. I came back home and wrote an ocean meditation.

(To be read out slowly on a safe place on a beach, with sufficient pauses between sentences and longer pauses between paragraphs for visualization. A sharing of what different people visualized can be invited at the end of the meditation.)

Close your eyes and bring your attention to the sound of the ocean. Listen keenly to the waves … if a thought arises, become aware of it and gently bring your attention back to the sound of the waves. Does each wave sound the same? Does the sound rise and fall? Listen to how each wave swells as it approaches the shore and then crashes on the sand. Listen to the brief silences between the waves. Listen to how the water crashes without rest and speaks in different voices, just as you hear the voices of different people who have gathered on the beach.

Now gently bring your awareness to your breath. Observe how you are breathing—deep or shallow. Through your left nostril, right nostril, or both. Can you smell the ocean air you are breathing? Is the wind changing your experience of breathing or the rhythm and force of your breath? Notice how your breath affects your whole body.

Imagine the ocean across its depth and expanse, and all the life forms which live in it. From the great whales to sea turtles to sharks to starfish to strange deep-sea creatures. Imagine the life-giving plankton all across the ocean providing this world most of its oxygen, which also grants you breath ... and the dolphin, the crab, the guitarfish and the seagull. Imagine this life breath swirling from the ocean into you, into other life forms and back ...

Breathe in deeply and breathe out.

Now imagine travelling back 400 million years, when life first emerged from the ocean and evolved into other forms, including yourself. If you witnessed this event, what would it have looked like? What would this beach have looked like? What if you could remember it deep in your cells, when you first emerged from the ocean as your long-ago animal ancestors? Imagine it now ...

Breathe in deeply and breathe out.

Now bring your attention into your body, its volume and what it feels like right now, into the blood flowing through you. Imagine the interstitial fluid which bathes all your cells. Know that it is identical to the ocean and carries a memory of your origin. Feel that through your entire body. In your head ... down your neck and chest ... your arms and hands, your stomach and abdomen ... thighs ... calves ... feet and toes. Feel the life-giving, life-sustaining ocean flowing inside you.

Gently bring your attention back to the sound of the waves. Listen to how they swell and crash. Listen to the silence between each wave. Listen to the various voices and noises of water. Let the ocean's sound permeate your mind and completely wash over you.

Urur Kuppam Beach
27 November 2020

Urur Kuppam Beach has turned into a battleground of garbage after Cyclone Nivar. More species of single-use plastic lie in thick, tangled mats above the high tide line than one could count. Oddities like bike helmets, stuffed toys, steel vessels and schoolbags are there too. A whole wooden sofa has washed ashore near the estuary. Trousers, saris, sanitary napkins and the elastic bands of undergarments. Then the residue of the pandemic—face masks of many styles and colours in their hundreds, sanitizer bottles and sachets, syringes, dirt- and salt-encrusted PPE gowns, baby ghost crabs crawling slowly amidst them.

The floods had flushed out a year's worth of garbage dumped into the Adyar River along various outfalls in the lower course. The longshore currents and winds had strewn it all across the beach. No, this is not the ocean coughing garbage back at us as is being explained in the news. This is the trash that was choking inland waterways.

Ragpickers move uneasily through the crowd to collect what they can from the shoreline. They know this happens after each storm.

The ocean is still muddy, though calm. Small spilling waves break on the intertidal, showing no swells further inside. Its winds are spent. The offshore waters are completely still. The ocean sounds like a forest brook flowing over smooth pebbles. Deep-sea pelagic birds which never venture onto land, except for breeding on unpeopled islands, were blown in by the strong winds and seen taking refuge in the marshes inland by birdwatchers. Among them are frigate birds, Arctic skuas, bridled terns and lesser noddies.

Artisanal fisherfolk classify the ocean-bed mosaic as *tharai, parai* and *seru*. Tharai are sandy habitats, shoals and other places where sand gathers due to currents. Parai is rock or reef underwater. Seru are slushy or clayey habitats where organic sediments have gathered over centuries. They are rich in prawns, crabs and other detritus eaters, which then attract large fish. Fishers have a vivid map in their minds of each area and its nature, and specific names for these places, especially serus, across their expanse of sea. The three different habitats have distinctly different species communities. Opposite Urur Kuppam, about half a kilometre southeast, is Kathala seru. This morning a number of boats had gone there to catch prawns and crabs.

But slushy bottoms are also very sticky surfaces nearshore. All of them come back with nets full of assorted trash, which they have to separate from their catch. Then they have to clean out their nets. 'It is going to take us the whole day to separate the nets and crabs from the trash,' Raghupathy, an elderly fisherman from Urur, tells me. The three-spotted crabs in his net are grabbing plastic cups and empty chips packets with both pincers, and he has to snatch them away from their grip.

The stormy sea has left deep furrows in the beach's berm, where the backwash is strong when the ocean surges. Riptides are locally called *vaangal*, sometimes described as ghost arms catching you by your ankles and dragging you into the depths. This whole beach is notorious for rips, especially when the weather is bad. Raghupathy tells a story of an old pandanus grove by the village where the ghosts of White people roam, those who had swum in this sea during rough weather.

Two mating carpenter bees fly as a couple, straight over the ocean. I see them with my binoculars till they vanish from its range, at least a hundred metres off coast. What was that about? Calm weather after a long while perhaps also means honeymooning bees over open sea.

Three days ago Cyclone Nivar, surging from the Bay of Bengal at 60 knots, had made its landfall near Chennai. Just before that I waded out into the street, which had become a shallow, slow-flowing stream, to walk a bit into the water's memory of Velachery, once a marshland, now almost entirely paved over. The streets going down from the railway station road opposite my apartment were flooded high up to an autorickshaw's wheel, the water stagnant upon impervious tarmac. The main streets flowed towards Pallikaranai. There was a power cut for much of the day.

Every storm takes me—and most Chennai people—back to the floods of December 2015 when most of Chennai went completely underwater for a few days.[1] I remember it well because I spent it in

---

[1] 'Report No. 4 of 2017 Performance Audit of Flood Management and Response in Chennai and its Suburban Areas', Government of Tamil Nadu.

something of an ark—at the administration office of Pathashaala, a school in rural Chengalpattu district. I took refuge with two other human colleagues, two barred wolf snakes, a common sand boa, colonies of black ants and procession ants spread over the walls, mud dauber wasps, sand wasps, field crickets, mole crickets, earthworms and other small creatures displaced by the floods. It was labelled a human-made catastrophe,[2] resulting from bad urban planning and abject disaster management in a report later released by the Comptroller and Auditor General of India.

Climate change may be measured in carbon, but its mother tongue is water.

Kaliveli Estuary
Villupuram
10 December 2020

Sempatagan or Bengal whiprays are arranged along the edges of the fibre boat docked at the backwaters near the fort. The morning sun is doubly strong for December, the skies having emptied in the past week. The sun's intensity is reflected by the lagoon from below. The tide is receding. There are pale islets of several hundred brown-headed gulls, Caspian and crested terns in the water, and the sandbars are stirring with flocks of plovers like moving laterite stones. Vijay, Vikas and I walk between the shore and the archaeological ruins of Alamparai Fort, having found a gap between lockdowns to make this visit. We are all masked up. Vikas can tell birds from their dark silhouettes, so familiar is he with their form and movement. Alamparai is a seventeenth-century Mughal seaport, captured by the French, then destroyed by the British, then destroyed again for good measure by the 2004

---

[2] Krupa Ge, *Rivers Remember* (New Delhi: Context, 2019).

tsunami. Now large portions of it lie under the backwaters and in the sea, establishing foundations for labyrinths of oyster reefs.

The Kaliveli wetland system starts at the estuary and moves inland, where there are islands of rich mangroves. Seagrass and oyster beds cover the estuary bottom and the salt marshes along its cheeks. The creek curves south, thins and runs 18 kilometres before reaching the vast Kaliveli Lake. Between December and March the wetlands host tens of thousands of waterfowl and shorebirds. On the estuary's sandbars, that magical mascot creature, the olive ridley sea turtle, nests in numbers—an average of 36,000 turtle hatchlings emerge from this stretch of beach. *Panavai* or sperm whale sightings in the ocean are common here, and they breed in the deep canyons offshore of the wetland, cetologist Karthik Ashok tells me. Fishermen speak of seeing them sleeping vertically when they go further into the ocean for tuna or king mackerel. Sperm whales have their blowholes almost near their foreheads, and often they rest by suspending themselves upright, letting their heads bob in the swells. I found a single whale vertebra on the beach, thick as a palm trunk, heavy as a shot put ball. It could be used as a stool. I sat on it for a while and watched a Jerdon's bushlark whistling atop a sand dune.

As this wetland stretches from sea to estuary, then to a long, curved creek and a large lagoon, its life, water and people gradually change down the salinity gradient. By the bar mouth are the relatively affluent seafaring villages of Alamparaikuppam and Azhagankuppam, to the north and south, many of them operating trawlers in harbours at Chennai and Puducherry. The ocean is richer near any large coastal wetland complex as sediment inflow makes the inshore waters proliferate with plankton, which then draws the entire marine food web. On the estuary behind the bar mouth the salt marshes spread across, as do the oyster beds. Large flocks of waders from painted storks to shanks, stilts and plovers can be seen here. Over a thousand women from twenty surrounding villages wade to pick oysters and collect their flesh in their *pari*s (straw baskets). Journalist Krishna Chaitanya through his interviews has found that coincidentally many of them are widows, having lost husbands to the illicit liquor business. They sell oysters and other bivalve meat to shrimp hatcheries, earning about 300 rupees a day after about eight hours in the water. The shells are dried and sold to lime kilns. Other men and women wade to catch prawns and crabs. There are groups of Irular here who catch lugworms from the mudflats.

In the middle of the estuary is Edaiyanthittu, a sandbar island of mangroves and halophytes (salt-loving and succulent plants), where the rare grey-tailed tattler is spotted every migration season, the only sighting of the bird in India, other than the Pulicat Lake. Further up is the Uppukali Creek, spreading into hectares of salt pans and supporting several thousand salt workers. The creek widens and spreads into a lake, where the water is mostly fresh. I sometimes avoid coming here during bird counts, for the tens of thousands of pintails, garganeys and other ducks which gather here and fly from one end of the waterbody to another make counting them the most arduous thing in the world. My eye aches the whole day after peering for hours through a spotting scope. Grazers and farmers depend on these waters.

In January 2021 public hearings are announced in the sea-facing villages of Azhagankuppam and Alamparaikuppam for two mechanized fishing harbours proposed in the estuary.[3] Along with them will come two training walls at the bar mouth, of 600 metres and 400 metres, and the creek is to be dredged to keep it accessible to 110 mechanized boats and 300 motorized boats. This harbour is a decades-long demand of the two seafaring villages, though it will be to the detriment of numerous other villages dependent on the waters of the estuary and the wetlands further in. Even though

[3] S.V. Krishna Chaitanya, 'Twin Fishing Harbours Cement Crisis in Kaliveli, Livelihoods at Stake', *New Indian Express*, 27 February 2022.

the project is only at the proposal stage, mechanized boats and trawlers are already parked over the estuary, smack over seagrass.

There are divisions and complex conflict zones along the salinity gradient of this creek and of any coastal wetland. The seafaring boat owners are rich and act as upper caste. Sustenance livelihoods like those of the women who pick oysters or lugworms in the wetlands are treated as lower caste.[4]

We wonder what we can do for this estuary to prevent it from being dredged and built over and stop the beach getting eroded away to the north of the training walls. Especially while the people living here are convinced that this is of no consequence compared to the wealth a harbour would bring. Or so go the stories told by the mechanized boat owners whose vessels and equipment cost over fifty lakh rupees to a crore.

We make a list of all the birds and other creatures we see at the creek. We take pictures of the large flocks of gulls and terns, the colonies of ibises, whimbrels, fiddler crabs, and the congregations of mud creepers and girdled horn snails on the rich tidal flats, which give us black slush stockings till our knees.

Kovalam Creek
6 January 2021

Who sent word to the oysters, limpets and barnacles that granite rocks were being dumped here? Standing on the second groyne of Kovalam Beach just a few inches behind the reaches of the cat-leaping waves, Rohith and I ponder the question. Far above in the sky pelicans and painted storks soar over the creek, lifted and lofted by arms of updraught winds, utterly unseeable to us. We are here to survey the catch and

---

[4] S.V. Krishna Chaitanya, '50-year-old TN Woman's Livelihood in "Troubled" Waters', *New Indian Express*, 2 August 2021.

bycatch brought in by the local fishers, to understand the life of this seabed. The beach is eroding here reportedly because of the shore temple at Mahabalipuram, made of gold-coloured granite, built in the seventh century, now blocking the longshore currents. Rows and rows of black granite are now being dumped upshore to trap the sand. On the groyne rocks there are rock-loving intertidal species which never occur on the mostly sandy beach habitats of the Coromandel Coast. But when rock does appear, albeit unnaturally, and the seawater feels it, associated life somehow makes its way there.

On the groyne's nose, attached to and living on these large granite boulders—mined from inland mountains—are rock oysters, covering almost entire surfaces of some rocks, then striped acorn barnacles, limpets, colonies of tiny littorina and turbo snails, and grey frog snails on the sea lettuce growing along the sides of the structure where water doesn't pummel but flows gently. In the low-tide zone where the sea laundry whacks the rocks, thick mats of chaetophora algae grow, on which dozens of mottled lightfoot crabs graze patiently. They crouch and grasp on algal turf when a wave slams over, then rise up to forage cautiously with their purple pincers during the lulls. Three pairs of lovers sit on the groyne's nose in each other's embrace, gazing at the sea. Many wandering glider dragonflies hover over their heads, as if bathing in their body heat.

I have an epiphanic wind of the seawater's pregnancy. Every curl and lapping invisibly brimming with eggs, larvae, zoea, cyprids, trochophores and other inscrutable planktonic and zooplanktonic potencies. Its seething, beating aliveness. Such that when the sea senses sand, senses salt marsh, senses rock or mangrove root or jetty plank, life climbs out in resonance, like a profound conversation.

After last year's cyclone quite a few granite rocks rolled off south onto the beach lying on the lower intertidal. On it I find an Anjuna sea anemone among the titan barnacles. The anemone is named after a beach in Goa. It is wide as a one-rupee coin. It had collected with its tentacles all kinds of debris—shell shards, algae, sand—and covered itself, hiding from eyes and heat till the tide rises. On the beach we find more of its cousin species—sea cauliflower drying, a red sea fan which has stuck itself on an ark shell, and bleached hard coral which has etched itself and merged into the shell of a virgin murex. The sea surges in and washes over us, drenching us from the shoulder downwards. Anemone begins to open.

Coral reefs, sea anemones, sea fans and other marine creatures are living architectures of friendship. Common to them all are zooxanthellae—alchemical, single-celled, photosynthetic plankton. They form friendships with polyps, anemones, sea slugs, clams, sponges and flatworms. They build coral reefs—complex structures of deep symbiosis—which have the highest biodiversity of any ecosystem on earth.

In the most uncertain of places, on rock stretches under constant flux and hardship, communities of life thrive through friendships forged by zooxanthellae. They challenge theories that propose competition and one-upmanship as the key to survival. They flout these rules. 'Competition makes sense only when we consider the unit of evolution to be the individual. When the focus shifts to the level of a group, cooperation is a better model, not only for surviving, but for thriving.'[5] What might a coral society be like in the human world? Is that a meaningful metaphor for us or is it too much zoomorphism? What might zooxanthellate politics be like? Beings which know cell-deep that 'all flourishing is mutual'. Who is to say what the 'self' is—asks coral, asks anemone. Who is to define its edges? Looking at and learning about these beings permeates and perforates one's own selfhood and lets one enter life's shore-like porosity.

[5] David Sloan Wilson, *This View of Life* (New York: Pantheon, 2019).

Urur Kuppam Beach
22 January 2021

Belt of Venus—
a hawkmoth flits
from headlight to headlight.

A new season begins for life on the coast. The first confident south wind blows—locally called *kacchan kaathu*—and the longshore current has reversed directions possibly till next monsoon. Longshore currents are those which flow like rivers near and along the shoreline. The berm of Urur Kuppam Beach smoothly segues into water in most places like a gentle ramp. Anchovies and silver biddies swim right at the waves, and one can see the glint of their silver bodies as the water curls. The intertidal zone is packed with ghost-crab homes, and there is a heap and collage of shells deposited near the estuary. Today many crabs have pushed out large mounds of sand, like they are shutting off their burrows with a massive lid. From what? Cold winds or mist? Are they all together doing some home remodelling? Olive ridley turtles have begun arriving to nest and their numbers ought to increase in the weeks to come. A ridley's hollow-eyed carcass lies at the Urur dockyard, shell cracked by impact and beak buried in sand. There are ghost-crab burrows all along its rotting flanks, its skin punctured and carved out at many spots by their scavenging. May less of them be struck by trawlers and get caught in ghost nets this year.

Goose barnacles grip almost everything which once floated and has been cast ashore this week. It seems that the ocean surface has a vast variety of footwear. Alcohol bottles have goose barnacles along their curves, like the scales on the back of a stegosaurus. Barnacle growth and density indicate how long they've been at sea. In May 2012 a man died for reasons unknown in the Tyrrhenian Sea, then floated out for a period of time, and later was washed ashore one of the beaches of Calabria in western Italy. Possibly for the first time, a forensic team used the deposition and growth rate of goose barnacles attached to his shoes and clothes to ascertain that he had been dead and floating for over three months.[6] It would be interesting to apply the same forensics to the trash on Urur Kuppam Beach and its warmer tropical waters. The findings may go something like this: Bata sandals—three weeks, Coca-Cola bottle—five weeks, milk packet—one week, bottom of refrigerator—eleven weeks, and so on.

Neetil is the word Chennai fisherfolk use for the long line in the sea which separates the murky waters of the wave-laden nearshore and the deep blue calmer offshore. It's a distinct meeting of colour fronts—which they call *edappu*—one clearly sees from the beach. Usually when the edappu is near the beach the weather is likely to be calm or sunny. But if it is far away or not to be seen, there is always overcast weather and a rough ocean. Travel through the ocean and one crosses many such neetils as the water's nature keeps shifting from changing and meeting currents, winds and bathymetries.

The mind mirrors the ocean.

Before I sit to write in the mornings I practise watching my breath—drawn from two-thirds ocean and one-third land—and the waves of turbidity and clarity which wash over me as I pay attention. As you practise attention, right at the beginning feelings and voices and images of one's immediate circumstances ebb and tide. Then maybe ten, twenty, thirty minutes later you cross a neetil. The mind becomes quieter, emotions emerge from deep within, sometimes raw and primal images arise that are outside logic. Light dances, long-ago memories may bubble up, the backbone may sway back and forth. Somewhere about here there

---

[6] P.A. Magni et al. 'Evaluation of the Floating Time of a Corpse Found in a Marine Environment Using the Barnacle *Lepas anatifera*', *Forensic Science International*, 2014.

is another neetil, a little elusive if you go searching for it. When you cross it you find your confined identities peel away. Ear canals may begin to hum. Eddies of thought only occasionally cross the expanse. The perceived boundaries between air and mind, land and body, inner and outer, self and other, may start becoming indiscernible. I try to enter this space before I write. It helps me open up and see the co-authorship of everything written together by sea, tree, insect, sun, barnacle and myself. A different kind of transrational, perceptual process operates. Anthropologist Tara Lumpkin's work shows how we can and we need to explore different layers of our consciousness to learn to coexist with other species and the rest of nature.[7] She has written about how most indigenous communities are polyphasic—they live in and value the importance of different states of perception beyond just the normal waking state. She calls this perceptual diversity. Like the ocean, our neetils are many.

The red bundles of tentacles—cirri—of the goose barnacles have shot out of their shell plates. They are unreeled when a wave crashes and rolls in, even in those barnacles stuck on rubber slippers and bottles many feet away from the intertidal. Attached like stickle bricks upon each other, older ones at the bottom and young ones at the top, the barnacles are all listening deeply to the ocean through each other and the glass and plastic on the sand. They know very well the vibrations of a wave crashing forward or washing back, and their cirri came out just in time when the water is near and they could filter feed from it.

Falling tide …
the howl of wind
in the turtle's skull.

---

7 See Tara Waters Lumpkin's work in voicesforbiodiversity.org.

North Chennai Toxic Tour

24 January 2021

The restless sea slaps over the rocks of groynes, which stick out into the ocean like spines of a city-sized stonefish. Surf flies onto the roads and into the tyres of large container lorries. Fibre boats float near the beachless coast and far-out cargo ships block chunks of the horizon. I have brought the Class 11 students of Abacus Montessori School on a tour through North Chennai, a watery landscape of long-time industrial violence. During the lockdown these students were part of a campaign to save the Pulicat Lagoon from a megaport proposed a few kilometres south of it, which would erase the second-largest brackish waterbody in India from our maps. The students wrote a letter to the chief minister and other authorities and got it signed by over 500 students from fifty different schools in Chennai. Some of them made powerful campaign art. We held a press conference around their campaign along with fisherwomen from Pulicat, with the corridors of the hall adorned with art exhorting people to save the Pulicat Lagoon. The media was moved, and the papers wrote strongly on the topic. The next day the district collector of Thiruvallur cancelled the public hearing planned for the port, putting a spanner in the works of Marine Infrastructure Development Private Limited, a branch of the Adani Group.[8] Today, with a little lift in the lockdown, the children can finally do a field trip to see this waterscape.

I am reminded of my own first toxic tour in North Chennai two years ago. I had sought out Nityanand Jayaraman (Nity) when I was just aspiring to be an active part of Chennai's socioecological

---

[8] Press release no. 24, 'The public hearing on 22 January 2021 for revised masterplan development of Kattupalli port has been postpone [sic],' District Collector Thiru. P. Ponniah, IAS, 19 January 2021. Issued by Government of Tamil Nadu.

issues. I heard him speak at my school first when I was eleven years old. He was (and still is) fighting Hindustan Unilever for its mercury poisoning of Kodaikanal's forests, lakes and factory workers.[9] My first exposure to corporate crime and the ways to stand up against it was through him.

As our bus crosses the Cooum River's estuary, the city begins to change. The residential and industrial landscapes begin to overlap in this place where the Kotralayar ('pouring wave river', referring to its torrential gushing during monsoons, Anglicized as Kosasthalayar) sprawls like a great, sinuous python. It flows rather linearly through its middle course, but enters Chennai and turns wild. It dilates and deflects north near Manali, where the petroleum flares from Chennai Petroleum Corporation Limited go on all day, and the air smells of mercaptans. Then the river widens into 10,000 acres of coastal wetlands and backwaters. It meets the sea at the Ennore Creek where smoke stacks spew black breath and the sky hisses with pylon cables—carrying 450 kilovolt amperes—on fields of fly ash.

The unsatisfied river then darts further north and opens out again at Karungali, where its estuary has been blocked by a port road, which the river blasts through every monsoon. The port authorities lay it over and over again. Then it finally flows into the vast Pulicat Lagoon. The sediment it has brought over the aeons has formed Kattupalli, a sand-barrier island, which shelters the river and wetlands from the Bay of Bengal and stands as the city's first barrier to cyclones.

In the human world, North Chennai has a dark and difficult history. 'Black town' was what the British called it. The people living here are largely made up of fishing, tribal, scheduled caste and working-class communities. Most live hand to mouth, dependent on land, sea, wetland or daily wage. The first colonial

[9] Nityanand Jayaraman, 'Unilever's Mercury Fever: Dumping Toxic Mercury in a Biodiversity Hotspot', *Conservation India*, 2011.

port, power stations and dirty factories were built here on the backwaters, not because the site is especially suited to these activities but because the people here could be financially and politically overpowered. The more historically privileged people lived in the south, and their travails were far less during the colonial period. But this legacy of oppression in the form of industrialization continues to date. Thirty-seven different red categories or highly polluting industries are present here now.[10] All built and expanded after India's independence. Each industries spreads its own flavour of pollution and disease. A fisher-activist from Kattukuppam, Ennore, Kumaresan Anna, not too long ago lost his son to osteosarcoma. This was possibly or rather most likely from benzene exposure from petrochemical industries. People around the Ennore Creek have a thousand times higher-than-normal cancer risk.[11] Chennai's largest landfill in Kodungaiyur is also here, sprawling over 270 acres, housing the garbage of its more 'well-to-do' residencies.

After our school van crosses over to North Chennai, we first stop at Nalla Thanni Odai Kuppam, a once-coastal village whose westernmost Amman Temple stands at the edge of the sea wall. The village's expanse lies far in the sea. The beach is nonexistent here. Prashanth and I ask the children to scan the landscape and tell the differences they see between the south they are familiar with, and the north, which almost all of them have entered for the first time. Prashanth is a friend and fellow activist at the Chennai

[10] See storyofennore.wordpress.com for 'The Story of Ennore: Reclaiming Rights, Livelihoods and Ecosystems Around the Ennore Creek in Tamil Nadu, India'.
[11] 'Expert Committee Report for Assessing the Damages by Fly Ash in the Ennore Backwaters', March 2022. Committee appointed by the National Green Tribunal, Southern Zone.

Climate Action Group. Then we explain longshore currents to them. These are slow ocean currents moving parallel to the coast, caused by the prevailing winds. Each day they transport metric tons of sand and sediment up or down the coast. Fish schools, turtles and other marine organisms use these currents to travel, migrate or move to breeding grounds.

Along the east coast of India, approximately between March and November, longshore currents move south to north. Between November and March they reverse their flow due to monsoon patterns. Coastal infrastructure like ports and harbours block the paths of these longshore currents, causing accumulation of sand to one side and coastal erosion on the other. The sea slowly eats up the shore every year.[12] To the south, the Chennai port and its towering cargo terminals resulted in the artificial accretion of Marina Beach—known as the second-longest urban beach in the world. To the north, the sea eats anywhere between 3–50 metres a year.[13]

Then we stop at the VNC Bridge over the Ennore Creek to see the Vallur and North Chennai thermal power stations. Some of the students are struck by smoke stacks creating clouds before our eyes. Long conveyor belts over the creek crackle as they carry black coal into the power station. I tell the story of the 2017 oil spill off the Ennore coast when an outbound ship from Kamarajar Port collided with an inbound oil tanker—20 tons of fuel oil spread along 35 kilometres of Chennai coast, our own Exxon Valdez. Dead olive ridley sea turtles kept washing ashore all week. For some fisherfolk at Ennore, this event marks a time before and the time after—when the seas would no longer yield them fish generously.

[12] Nithu Raj et al., 'Estuarine Shoreline Change Analysis along the Ennore River Mouth, South East Coast of India, Using Digital Shoreline Analysis System', *Geodesy and Geodynamics*, 2019.

[13] U. Tejonmayam, 'Over 30% Chennai Coast Shrinking Due to Sea Erosion: Study', *Times of India*, 27 July 2018.

Our last stop is at the resettled village of Kattupalli Kuppam, where we listen to Yashodhamma speak to us. In her sixties now, she is a fisherwoman and a fierce activist against the privatization of her home coast. Most of these women have been twice evicted from their original beaches due to erosion and industrialization and are now on the verge of a third eviction for the new port. They acknowledge these children who campaigned for their lands and waters (something these women have been doing for decades) and then go on to tell of their struggles and resistance towards the port expansion.

The students are meeting the realities of industrial violence, eviction and community resistance for the first time. One of them, Maya, wrote an account for the school about the interaction with Yashodhamma and the trip to Kattupalli.

> She told us of the plight of the fishing communities residing there who no longer had access to the sea. She gave us a first-hand account of problems that a port could cause having seen the impact from the earlier ones. She also told us that the promises of politicians were futile, reiterating that only citizens could cause real change. She thanked us for our contribution to the project and was elated that young students were taking action. We also met a few other women who resided in the same area and heard their stories too. The relocation of local residents not only caused monetary problems, but psychological ones too, as fisherfolk felt a strong connection to the ocean.

Adyar Estuary
27 January 2021

Palayam Anna teaches me the wind system of Chennai's artisanal fisherfolk. I take notes in a pocket notebook. Wind straight from the sea is *ner eeran*, the northeast wind is *vaadai eeran* and the southeast is *kacchaan eeran*. These are good winds, he tells me. One can start to sea at night or early morning, and they will bring you back safely to the shore. I try to recall what he taught me and mix up the nine wind names and their directions, and he stresses, slightly annoyed, that I must listen more carefully. Palayam Anna is a fisher elder from Urur Kuppam, and along with Nity is among my greatest teachers of the coast and ocean, after the coast and ocean themselves. His uncles and other elders, he tells me, would be quick to give a hard whack to young boys who mixed up their wind names on the kattumaram.

I provoke a conversation about whether it is possible to teach without harming the learner, and what fear does to the learning atmosphere and the learner. Palayam Anna complicates my educational values. He calls me a 'land teacher' and says that my profession is on an unmoving and highly predictable surface, with little weather action happening, which allows me to have debates like these. He is, however, a 'sea teacher' and in his vast, fluid medium the conditions are constantly changing and highly unpredictable. One must attach one's fears to the right things, and firmly so. Speaking the right names can mean life or death. To learn at sea requires much more discipline and alertness than learning on land.

He left me thinking about how systems of learning would be different in an ocean world or if we were humanoid marine creatures. Or for young whales, dolphins and fish themselves.

Whether I remember any of the winds or not, Palayam Anna tells me that I should know for sure the *kun vaadai* even if I never went to sea. These are what he calls 'vicious' winds from the north–northwest, which indicates that a bad storm is about to strike soon, swirling in the nearshore. I try to correlate what he is saying with my understanding of geography and the coriolis effect. In the Bay of Bengal, and the whole of the northern hemisphere, cyclones spin anticlockwise. When their centres are far away from the shore, winds blow from the north (*nedum vaadai*). If they are worryingly nearshore, then their spinning arms curl around the coast and blow from the northwest.

A strange event occurred during the Indian Ocean tsunami of 2004, which drew much public attention. I was eight years old at the time and was sitting in a Carnatic music class at the teacher's home when her husband ran in and said the sea was rushing into land. All the children were sent home. In the news I saw the massive waves hitting the shores of Sumatra, Indonesia, the Andaman Islands, then Cuddalore, Nagapattinam and Chennai. Fisherfolk along the southeastern coast of India say that the ocean has drastically changed after the tsunami, that it has become more unpredictable and wayward than it was before the disaster. The tsunami is also constantly used as a reference event in their lives and a turning point, emotional as well as oceanological, just as historians might use before and after the Common Era.

Among the worst affected places were the Andaman and Nicobar Islands, where giant waves pummelled down whole horizons of coastal infrastructure. What was also widely documented during this time was how the native tribes of the islands—the Sentinelese, Jarawa, Nicobarese and others—left the shores much before the disaster and moved to the hills.[14] The great Indian linguist G.N. Devy, who conducted the People's Linguistic Survey of India, says, 'Their languages have words which let them sense the different textures of waves. As the tribes moved uphill, they supposedly told people, "The ocean is angry with us. We must move up in repentance."'[15]

I flipped through a dictionary of Car-Nicobarese by Rev. G. Whitehead to find some of these words.[16] The languages of these tribes are astonishingly ocean-centric. For instance, the direction system of the Nicobarese employs 'lö' and 'il' (affixes denoting

[14] Neelesh Misra, 'Stone Age Cultures Survive Tsunami Waves', NBC News, 5 January 2005.

[15] See 'War of the Tongues', G.N. Devy in Conversation with Payal Purie, Algebra Conversations, YouTube.

[16] G. Whitehead, *Dictionary of the Car-Nicobarese Language* (American Baptist Mission Press, 1925).

left when facing the ocean) and 'ti' and 'tö' (affixes denoting right when facing the ocean). Other ocean words I sifted from Whitehead's book are:

Ank: the call of the white-belled sea eagle

Èl–mai: waves of a stormy sea

Tö–lön–lö: something cast up by the waves

Hom pa–nam: to assuage the winds or waves by witchcraft

Ka–lūöng: to be calm or still (of wind or waves)

Pa–rë–nya: the lapping of water due to a breeze

Kui–mai: the horizon over the sea

Vi–ni–il–kuö pa–nam: a devil-scarer or a charm to banish stormy weather

Pa–sā–va: to be rippled

Pā–vi: to be cast adrift

Pā–va: to beat gently (of waves)

Nyô–nyi: sands drying out at low tide

Fa–tah–tö–re: to strike against one another (of wind or waves)

Ka–eò–ngö: to be tossed about by the sea

I have come across such ocean-centric speech among the artisanal fisherfolk of Ramanathapuram as well, whose livelihoods in the Palk Bay are strongly threatened by intensive mechanized fishing. Spatial references are always with respect to the ocean. *Villangu* is in the direction of the ocean, *karai* is towards shore, *kondal* and *kachaan* are to the left and right respectively while looking at the ocean. They have their own system of winds too, with slightly different names than those of Chennai fishers.[17]

Among Japanese artisanal fishers there is the concept of *kaze o kanjru* (feel the wind), which is to be in sensorial touch with the sea's conditions, wind force and directions as keenly as possible. The

[17] Shyam S. Salim and Monalisha S., 'Indigenous Traditional Ecological Knowledge of Tamil Nadu Fisherfolks: To Combat the Impact of Climate and Weather Variability', *Indian Journal of Traditional Knowledge*, 2019.

local names for winds and their distinct characters are numerous, differing on different coasts. In the fishing communities of Gomso Bay, South Korea, time is deeply tidal. Most clocks and calendars around the world follow the luni-solar movements. But the days of the Gomso fisherfolk have alternate names describing daily tidal changes, and their calendar follows a fifteen-day cycle—tide times stretching between a neap tide (*jogeum*) and spring tide (*sari*). They live and plan in spans and shoreline rhythms of six hours, twelve hours, twenty-four hours and fifty minutes.[18]

The African isiXhosa word for ocean is *uLwandle*. It means 'we are, because the ocean is'. The word falls in the same noun class as *ubuntu* in African Nguni languages. The ocean is not a singular thing nor an object which can be spoken of apart from the speaker.[19]

Pulicat Lagoon
1 February 2021

Around midway by the Pulicat barrier island we turn off the boat and float in the lagoon to see four golden jackals. Rohith, Vijay, Yuvaraj and I are on a boat searching for a specific creature which has been sighted swimming in the waters of the lagoon over the past few months. The jackals emerge onto the shore from a forest of prosopis and palmyra. Two wade ankle-deep in the lagoon, one of them splashes something with its paw and picks up a large crab carapace in its jaws and trots to the sand. When the boat's engine goes silent, all of them stand uneasy, the hackles on their black backs slightly erect and ears on alert. They watch us with side glances as they walk along the island, as if to hide the fact that they are keenly noting us. Jackals

---

[18] Sook-Jeong Jo, 'Tide and Time: Korean Fishermen's Traditional Knowledge of Multtae in Gomso Bay', *International Journal of Intangible Heritage*, 2018.

[19] See the illustrated short film, *Lalela uLwandle*, Empatheatre, YouTube.

are good at that. We watch three of them scan the water further while one stands yawning and bored, not really interested in looking for a meal.

Yuvaraj, our boatman and a fisherman from Pulicat, calls out a marsh harrier in the distance without binoculars. It is quartering over an exposed mudflat, sending a flock of turnstones rolling into the air, making it look almost like a bubble of smoke in the distance. Then he guesses a great thick-knee with only the tip of its bill peeking out around a sandbar's curve. His capacity to identify birds with the slightest of their features amazes me.

Morning to the lagoon's north is made of horizon-long lines of shimmering painted storks and greater flamingos. We park our boat at Kuruvi Thittu, a small island with old-growth Avicennia mangroves whose long snorkel roots have nearly barricaded entry into it like a fence of arrows. A flock of 300 golden plovers wheeze and sneeze over us, rising from the island and folding deeper into the lagoon. Between the mangroves are large tide pools made by the upward-turned roots. Small, balloon-like egg sacs of red lugworms are floating in them. Here and there is a sudden burst of marsh bluetail damselflies, insects which can breed in brackish water pools.

Pulicat is the Anglicized version of the name Pazhaverkadu, which in Tamil means an ancient jungle of roots. Human and colonial history is vivid here, from the Arabic to the Dutch to the British who used and shaped this lagoon as a port for trading and slavery. The original architects of this place, as its true name tells, are the grey mangroves (*Avicennia marina*). Each mangrove species has developed its own slow coastal sorcery. Avicennia is a coast-builder. To adapt to tidal variation, it puts out snorkel roots or pneumatophores which stick upwards out of the ground like thin, spindly, springy fingers, 10–20 centimetres long. During high tide the root tips can still breathe; during low tide the trees have around them a bed of arrows. In the pools around these snorkel-root beds baby fish schools are plenty. Fish from the seas migrate here into the Avicennia nurseries sheltered from the waves and velocities of the bay, breed and return. But root beds over time trap sediment, making land surface. New seeds fall, new root beds form, sediment is trapped, more coast is made. Avicennia are the true architects of this whole landscape, making and stabilizing the riverbanks, edges and sandbars of the lagoon. This wetland is also a bird sanctuary and a garland of artisanal fishing villages.

Less than 1.5 kilometres south of Pulicat, the Adani Group has proposed a megaport expansion of 6,111 acres of the existing L&T shipping yard of 330 acres on the Kattupalli sand barrier island. Of these, 3,000 acres will be on the ocean, the Kotralayar River and coastal wetlands, made by dumping river sand and assembling land out of water.[20] In its environment impact assessment report the company shows a flood map—or rather water's memory—of the places which will be several feet underwater due to the megaport. These are places where fisher communities live now. But the biggest threat it poses is that it will merge the Pulicat Lagoon and the northern part of the Kotralayar River with the sea.[21] Every year during the northeast monsoon, cyclones rage towards the city from the northeast. At the northeastern corner of Chennai are the Ennore–Pulicat wetlands, the city's most important climate barrier and flood catchment area. These sandbars, shoals, mangroves, sand dunes and salt marshes take the first hit from each cyclone and buffer its force.[22] In its absence, I think about how the storms will hit this climate-vulnerable region—with the same force felt in places several dozen kilometres inside the ocean.

[20] Bharathi S.P., 'Why Pulicat Residents Are Opposing Adani Port Expansion in TN', *News Minute*, 28 January 2021.
[21] Meena Menon, '"Adani Will Gain": Why TN Residents Are Protesting Plan to Reduce Pulicat Sanctuary Buffer Zone', *Scroll*, 1 April 2021.
[22] Saradha Natarajan and Benisha B.M., '"Sea Belongs to Fishermen": Locals Against Adani Port Expansion', *Quint*, 11 March 2021.

Taking off from Kuruvi Thittu we search for the smooth-coated otter, which has been sighted a few times by the local fisherfolk—for the first time at Pulicat—popping up and diving into the waters like a darter bird or sometimes at the edge of a sandbar crunching fish, as Yuvaraj describes it. We look for other species and other stories which will keep this waterscape alive in the public's mind and help in stopping the construction of the port.

Pazhayar/Kollidam Estuary Cuddalore
27 February 2021

The noon sea is greenish-blue, almost like a polynet envelope. The bright sun from the clear sky throws open the shallow ocean bed's colours and surfaces, mostly sandy with some patches of slightly darker sediment rich in mud and clay. Kollidam, the northernmost distributary of the Kaveri River, joins the Bay of Bengal about a kilometre north of Pazhayar. While driving on the bridge over its estuary I stop and see two wire-tailed swallows hawking insects over the mangrove clouds. Sand is usually beige, but during this time of the day the silver grains in it sparkle, and swirls of black magnetite stand out on the beach. On the foreshore, a few metres from the high-tide line, a sand wasp (*Bembix sp.*) is digging a burrow, dribbling a spray of fine silt at super speed. The heat feels like a trap-jaw on my temples.

Boats are returning to the dockyard south of the estuary. There are humpback dolphins in ones and twos, porpoising close to the shore in the same direction, their silver backs and dorsal fins catching the sun like aluminium foil.

My friend and marine biologist Rahul Muralidharan studies the interactions between humpback dolphins and fisherfolk near the Gulf of Mannar, a militarized seascape between India and Sri Lanka.[23] The trawling lobby is notorious there and the use of banned nets like purse seines and pair trawls is common. These are stressed, overfished waters where little fish is left behind for both artisanal fishers and dolphins. Humpback dolphins, which live in shallow nearshore waters (locally called *ongil*), frequently rip fishing nets and take the catch. Humans retaliate, sometimes kill.[24] Dolphin and human have become enemies here, competing and fighting for the same scarce resource—fish. Alexis Pauline Gumbs calls this dolphin behaviour of 'stealing fish' as 'levying tax' for the use of their territory.[25] I love this perspective of cetacean resistance to human appropriation.

In Pazhayar, humpback dolphins are not hated, though there is a harbour beyond these artisanal villages where trawling happens, stressing out the seas. I speak with Kasiraja at his home while returning from the beach; he has been an artisanal fisherman for the past fifty years and is an elder in the hamlet. 'We are happy to see them. They tell us if fish have come near the shore and we too watch their movements carefully,' he says of them with seeming fondness. I wonder if he takes me for a forest department official but his story feels genuine. Here, dolphins are called *vedan* (translating to 'hunter' in Tamil) or sometimes *sami sura* (god shark). 'They rarely bite our nets. If they get caught by accident we release them immediately,' he says, describing how the two species, human and cetacean, seem to understand each other's boundaries. 'Groups of vedan lead us to large *mappu* [fish schools] further in the sea. At other times they follow our boats.' He mentions that as a child he had seen so many more of them and his elders had taught him that they are not to be harmed.

[23] Rahul Muralidharan, 'Entangled Lives of Dolphins and Fishers', *Journal of Threatened Taxa*, 2018.

[24] Dipani Sutaria et al., 'Humpback Dolphins (Genus Sousa) in India: An Overview of Status and Conservation Issues', *Advances in Marine Biology*, 2015.

[25] Alexis Pauline Gumbs, *Undrowned* (AK Press, 2020).

I walk back to the beach from his house and see nine dolphins, less than 50 feet from the beach, prancing in the shallow, warm waters for a long time before swimming north. Three Caspian terns with red beaks fly along with them, their calls subdued in the wind.

Poigainallur
Nagapattinam
28 February 2021[26]

The flipper tracks of an olive ridley turtle break out of the tide line and veer onto the beach. She has nested somewhere nearby and returned to the sea. We try to look for where she might have dug her nest and laid her eggs. Adjacent to the beach are large sand dunes, some 40 feet high. I am walking around in the coastal village of Poigainallur, in the Nagapattinam district of Tamil Nadu, with Isaiyamudhan—a friend and fisherman from the neighbouring district of Mayiladuthurai. I met him through my mother who is helping his family with legal matters.

Poigai is the Tamil word for freshwater pond, and this village is not far away from the famous shrine of Velankanni. It's the last day of a road trip my mother and I undertook along the central section of the coast of Tamil Nadu, visiting fishing villages, talking to people, documenting and understanding the various coastal ecologies along this stretch. I was also searching specifically for intact, well-developed coastal sand-dune ecosystems—perhaps the most fragile and threatened shore habitats in India.

The previous day I had visited some parts of Parangipettai, a township north of the Vellar River's estuary in Cuddalore. A

---

[26] A version of this passage first appeared in the June 2021 issue of *Current Conservation*.

place known for its massive coastal dunes in the past, we found that their numbers had significantly reduced, being flattened, built upon and quarried away in bullock carts in front of our eyes. On our way to Poigainallur, we cross a number of tsunami memorials dedicated to fisherpeople who have lost their lives.

Isaiyamudhan tells me a moving story about Poigainallur and the neighbouring villages, a story to which the land bears witness. The local fishers and farmers had for a long time a deep understanding of how indispensable sand dunes were for water security as well as for their lives. Dunescapes take centuries to form. Sand is slow water, a patient fluid, which is moved, shaped and folded by wind, waves and vegetation. It flows over the years and with the seasons, like a current in deep time. The villagers had the practice of sticking palm fronds on nascent sand heaps in the path of the wind to help their growth by trapping other windblown sand particles. They also buried palmyra seeds in them, which preserved the dunes as they grew. This practice, thought to have been prevalent once, faded over time as the dunes grew large and the ecosystem services they provided were taken for granted.

A white-bellied sea eagle soars above us, its body cloud-white and wings coal-black. We see its gliding form through the gaps of palmyra fronds. It stops circling, buckles its wings and suddenly crossbows towards the sea, deft of flight, like a giant arrowhead of flint and granite. Till the crowns of palms block our sight.

When the Indian Ocean tsunami ravaged the eastern coast of India, fisher communities bore the brunt of the disaster. Isaiyamudhan tells me that the force of the waves brought the ocean several kilometres inland along these districts. The toll it took on women's lives was significantly higher compared to men. They were homebound, stuck with children and their chores, and couldn't escape from the impact zone fast enough. Their long hair caught in the thorny prosopis bushes, which have invaded shore habitats, and several women drowned.

In many areas, people observed that sand dunes greatly buffered the impact of the waves and protected them, and villages where houses were located on the dune crests were largely unaffected. Dunes protected coastal communities during this time; in some places even better than mangroves. The importance of these dune systems became clear, and since these habitats took a beating from the pummelling waves, the old practice of growing and preserving dunes was revived for a time.

A dune lives a slow life, one which is difficult to observe within a single human lifetime. Small vegetation traps windblown sand grains formed by wave action, which slowly accumulate and are stabilized by dune-binder plants like *spinifex, ipomoea* and *fimbristylis*. They are also known to help dunes recover after storms and strong winds. As a dune grows, it shelters its landward side from strong, salt-laden winds, allowing shrubs and trees to take root. Palmyra grows. Then ficus seeds dispersed by birds grow on the palmyras and slowly become large trees. Over time, a dune valley grows into a forest.

The most miraculous thing about coastal dunes is their effect on the local hydrology. In the village of Kallar, I approach some fishermen playing cards under the shade of their boat after spending the morning out at sea. They testify that the *manal medu* (sand dune) 'created' freshwater and point to a hand pump on their beach, located just 30 feet away from the high-tide line. I jockey its handle and taste the water. Although a little muddy, it is fresh. They also mention that many dunes here were blown down during the Gaja cyclone in 2018–19, but thanks to the protection the dunes offered, their village was largely untouched. At low tide, I spot hundreds of wedge clams (*Donax cuneatus*) popping out of the sand to filter feed from the waves, while crows and egrets attempt to nab their soft, jelly-like bodies before they clam shut and burrow back into the sand.

A dunescape acts like a massive percolation chamber and is an extraordinary rainwater harvesting system. Where they are well-developed, the water table is above or at sea level. The undulating sand structures, when saturated with water, form a shield below the ground, blocking seawater intrusion. Swales— small and sometimes perennial freshwater ponds—form on the landward side of dunes. And within these I discover swimming tadpoles, dragonfly nymphs, whirligig beetles, diving beetles and water scorpions—freshwater life forms thriving just 200–300 feet away from sea. I see the fresh pellets of a black-naped hare near one pond. Fan-throated lizards, flicking their blue dewlaps, scamper into the spider-like roots of pandanus plants growing around the pools. Jamun, alangium, cashew, bamboo and some large banyans are among the trees growing in the place. Behind the last line of dunes—the largest of them with crests over 30–40 feet high—villagers are cultivating paddy, now nearly ready for harvest.

'That new temple in the next village was built with sand quarried from here,' a woman from Kallar says. According to most of the fisherfolk I speak to, coastal dunes are severely threatened by construction companies, contractors and increasing demand in urban areas for building infrastructure. It is sometimes the locals themselves who mine sand on their bullock carts and sell it, due to unemployment and for access to quick money.

In Tamil Nadu, coastal dunes are often classified as *poramboke* land by the revenue department. Historically, poramboke in Tamil meant land reserved for shared community use—this included waterbodies, grazing land, sand dunes, riverbanks and mudflats. These were habitats which provided essential ecosystem services and could not be privately owned or used for agriculture and construction. With the advent of colonialism in India, this word became twisted to mean 'wasteland' because common land could not pay the Raj tax. For the sake of tax these common and crucial landscapes were diverted for colonial infrastructure and private property. The word even became pejorative, although there is a movement now to revive and reinvoke its original meaning and significance.[27]

It was an interesting exercise to trace how the idea of 'wasteland' came to be and follow its roots into the colonial period. I have been able to collect over 140 words for 'land' in Tamil from sources including Sangam literature and people's accounts. Each word evokes land in its ecological significance, cultural values and practices, and poetic contexts, where land has its own agency, animacy and seasonality. But not one of them describes land as 'waste' or 'useless'. One of the first instances where the idea of 'wasteland' is systematized is in the English philosopher John Locke's theory of property, which found its way into Indian law and governance as far back as the 1790s. Locke said, 'Provisions ... produced by ... one acre of inclosed and cultivated land are ten times more than those ... lying in common' and 'land that is left wholly to nature, that hath no improvement of pasturage, tillage or planting, is called, as indeed it is, waste'.[28] The colonial vision of land completely disregarded ecological truths and the living communities which depended on them. Woefully, this vision has been carried forward by our governance systems and policies to date.

Dusk is near and the tide is coming in. I sit on spinifex grass running over a sand slope to watch the elusive Bibron's skink foraging, moving in fits and starts, flicking its tiny tongue among the spines of grass, abruptly orange in brown and green. The beach reptile's little footmarks from its stuttering crawl are almost immediately covered up by fine silt blown by the gentlest wind.

[27] Listen to 'Poramboke Paadal', sung by T.M. Krishna, written by Kaber Vasuki, conceptualized by Nityanand Jayaraman.

[28] John Locke, *The Second Treatise of Government* (Hackett Publishing Co., 1980).

Neelankarai Beach

2 March 2021

Low-tide sea. Ghost crabs scurry to take cover in the waves. Washed-up driftwood is twice as heavy with water and a tessellation of goose barnacles—travel stamps of a vagabond branch.

I began visiting this beach two years ago when my music teacher S. Balakrishnan passed on some of his students to me, one of whom lives nearby. I would time my classes near low tide, visit the beach and then go to teach. Later, at the request of the student's parents to not get their whole house sandy and wet, I would finish class and then come to the beach. I began learning the recorder (a classical European woodwind instrument) from my teacher when I was four years old. In 2018 he discovered a tumour on his leg, near his right big toe. I could seldom keep up with his stride, and even at seventy he would only take the bus. He would ask me to do the same if I picked up my phone to book a cab and tell me not to be a lazy young lad. His health deteriorated as the cancer spread. He asked me to start teaching music and passed on his students to me. In January 2019, I saw him, eyes closed, emaciated, on a bamboo stretcher. He had passed away after bringing the blessing of music into hundreds of lives. The last songs we played together when he was weeks close to death were Telemann's 'Concerto in E Minor for Flute and Recorder' and Bach's 'Brandenburg Concerto No. 4'. His breath rasped through the scant space in his lungs, in between the metastasizing disease. But the flute and recorder sounded and yielded to his will.

Later in 2019, in November, my younger sister, Yazhini, passed away. I took her to the hospital for stomach cramps, and the allopathic medicines began to have side effects. Her kidneys shut down soon, and in about a month she suffered multiple organ failure. I did her last rites and dissolved her ashes in the sea. To hold my deeply loved sibling as grey bone powder in my hands and to leave her on the waves was the most grief-filled experience of my life. In the hospital, with tubes running into her, she would ask me what species of moths were perched on the wall each day. I'd tell her the scientific names, as few moths had common names. That would perplex her a bit, trying to pronounce them. Then she'd ask their Tamil names, which are nonexistent. During her funeral moths thronged the walls of my home and the ice box and garlands. It was unexplainable. This continued for several days later, in numbers and species I have never seen locally.

When I come to the beach I often think of her, the seashore washes back memories of her, sorrowful and happy. When I see many moths on a wall her memories fly back to me. I imagine her spirit having mixed with the ocean.

As one walks away from the public beach at Neelankarai, the scene of the intertidal abruptly changes. When each incoming wave bounces beyond the tide line, a colony of cuneate wedge clams pops out of the sand in unison. On certain stretches of the coast suddenly their numbers are large; they disappear days later and reappear in other places—a mystery I have not unravelled. Each clam is a pair of shells packing featureless protoplasm. A hard calcium sandwich of cloudy white jelly. They pull out their two siphons to filter feed off the water, and when the wave retreats they quickly bury themselves back into the sand with their spade-like muscular foot. If you dug with both hands and shovelled up a chunk of wet beach sand, it would have not less than a dozen clams in it. These are creatures of the intertidal, adapted entirely for the land, sea, sand, air edge. Fisherfolk call them *matti* and sometimes come at low tide to collect them. They are not commercially sold, but are pried open, smoked and eaten at home, or used as bait.

The clams are in trouble if they don't flip themselves over speedily and bury themselves before the wetness of the wave recedes. The next wave might fling the bivalve out onto the beach, where it might dry out and die. Crows and egrets also haunt the shoreline, picking off late diggers. The birds try to grab them by their jelly limbs and shake them off before they can shut themselves. I've seen crows sometimes carry closed clams away from the water, wait, lose patience and fly back to the tide again. Those which are caught are slurped down. A hinged pair of shells is left shining open on the beach. Over time they become sand again, or the shell of clam or some other creature.

Among the clams lives another intertidal species. It looks like a large, nacreous bullet oozing out muscle. It also emerges out of the sand as the tide falls. Its tracks on the shore are distinct little trenches as if they were drawn by human fingers, as it doesn't dig but simply crawls into the sand at a shallow angle. The olive sea snail (*Oliva sp.*) is carnivorous, and lives buried amidst its prey, hunting them only above ground. An unburied snail uses the lapse between waves to feel around for an exposed clam. If it does find one nearby, it wraps its large muscular foot onto the appendages of the clam, which squirts water and fights to shut itself. Slime gropes slime, squirming together, creating a whirlpool of wet silt. The snail tries to drag the bivalve under the sand before the next wave hits them. But more often than not, an incoming wave tosses them apart and the olive searches for a clam foot all over again during the next wave intermission. And so life unfolds at the intertidal in incessant uncertainty. The olive sea snail is commonly known as *kovanchi* among the fisherfolk, a name alluding to its similarity in shape to the ivy gourd or *kovakkai*.

They can be found in different colours, from several morphs of white to ivory to sandal to grey-black. There is a study on how the colours of intertidal gastropods are temperature-dependent.[29] The darker-coloured ones absorb more heat through their shells

[29] Osamu Miura et al., 'Temperature-related Diversity of Shell Colour in the Intertidal Gastropod Batillaria', *Journal of Molluscan Studies*, 2019.

and are at greater risk of drying out than the lighter-coloured ones. On these beaches, white olives are the most common, more in number than sandal and grey. But as Chennai's summers get warmer by the year, white olives may dominate as other colours dry out.

In all of India, this part of the southeastern shoreline has the smallest tidal range. About 4–5 feet is the vertical difference between high tide and low tide. Two arms' length, literally. And this slow ecological drama involving wedge clams, olive sea snails, ghost crabs, purse crabs, mole crabs, ribbon bullia snails, lugworms, hermit crabs, periwinkles, whelks, shorebirds and so many other creatures occurs on this lean strip of sand— appearing and vanishing between the tide lines.

Urur Kuppam Beach
23 March 2021[30]

I've come to the beach at low tide. Thin brushstroke clouds are dissipating in the March sky. It is one of those days at Urur Kuppam Beach when adult horn-eyed ghost crabs gang up and stroll together on the surf. Adult ghost crabs are strictly night creatures, but once in a while they do this day patrol. Here and there a few slow down, stop to inspect debris or pick up a morsel from the sand. Silhouettes of Caspian terns send some sprinting sideways into the waves, and they tiptoe out later. There are about thirty of them in a loose group, all strolling at the same pace and in the same direction, and there are one or two more groups further ahead. Vikas and I follow the patrol along the shore. He lugs a massive telephoto lens under the midday sun, pointing it

[30] A version of this passage first appeared as 'Sand Scavengers: Ghost Crabs in the Intertidal Zone of Chennai's Urur Kuppam Beach' in the September 2021 issue of *RoundGlass Sustain*.

every few minutes at some bird in the sky or far into the ocean, like a ghost crab's eye.

Down this beach, past the fishing hamlets of Urur Kuppam and Olcott Kuppam, and beyond the forest department's turtle hatchery, is the Adyar River's estuary. The crab troop begins to scatter as we approach the river, where egrets and sand plovers are standing in the swash zone (where waves break) foraging wedge clams. Two excavators stand there to dig up the river mouth from getting blocked. Adyar comes from the Tamil word *adaippu,* which means getting blocked, which it does naturally often due to its sandy bed. But many skyscrapers, including the Leela Palace hotel, have been built within its creek. If the river is allowed to behave naturally, many buildings will have the Adyar inside them.

A dark brown mudflat stretches from the cheeks of the river's mouth where another ghost crab species shows up—red ghost crabs. The young ones are drab brown, but the adults are red as cherries. They are not beachcombers like their sibling species, but sand-sifters. They live away from the wave action, forage organic matter from the mud and leave a trail of near-polygonal chunks of processed soil outside their burrows. Reds are shy and seldom venture more than a metre from their lairs. At the estuary, the river flows into the sea carrying the shadows of the city. Waves foam at its mouth, and the water is greenish today, smelling of algal broth and sewage outfalls. A broken bridge juts over it. A bunch of college youth stand at its precipice and take selfies.

The bridge stands in three disjunct pieces as a testimony to the temperament of this river–sea junction. In the 1970s, a flood shattered most of the bridge and sent its fragments into the Bay of Bengal. Yet another project to build a road over the estuary is being considered that will displace the fishing hamlets that have been here for centuries. Under the bridge's southern fragment, fields of fiddler crabs appear as the tide falls. Ring-legged fiddlers

(*Austruca annulipes*) and fluorescent-blue variegated fiddlers (*Austruca variegate*) wave the larger of their claws vehemently over their heads. I keep silent for a while and can hear the air murmur with pincer sounds. The Students' Sea Turtle Conservation Network records on average over 200 olive ridley sea turtles nesting on this beach in winter.

The horn-eyed ghost crabs are now loosely dispersed; many have begun their return stroll away from the river towards the beach again. Two have found a beached strophidon eel carcass and are cleaning its spine out disc by disc. Another one crouches among the waves, feigning dead for a while—eye stalks drooping, legs collapsed—but then jumps to life and zips off into the sea when we approach.

A fisherman and his young boy from Pattinapakkam across the estuary walk past us with a bag of ghost crabs they have just dug out from the intertidal zone. They tell us that they are taking it for breastfeeding mothers in their hamlet. *Paa nandu, karuvaali nandu, kuzhi nandu* are some local Tamil names for ghost crabs. Their meat is used as fish bait; it is also strewn on the shore at low tide to lure lugworms to the surface. In the Tokelau islands fisherfolk say that these creatures are good weather forecasters.[31] If a crab chucks sand at a distance from its burrow as it digs then the weather will be calm. If it throws sand cautiously close or stays inside its burrow all day then strong winds, rain or tidal surges can be expected.

To me, the life and physiology of ghost crabs is sheer science fiction and the stuff of mythology. They are everyday company, yet their utter 'otherworldliness' has a pincer-hold over me. Ghost crabs are coastal chimaeras, edge-of-the-world denizens, living on the foggy cusp of land, sea, sand and sky. They have lungs and gills to breathe in both air and water. They can hear and drink with the setae (hair) on their feet, speak in pincer

[31] Elders from Atafu Atoll 2012.

signs, stridulations and gut rumbles. They have panoramic 360-degree vision, and the cylindrical retinas atop their eerie, ghostly periscope eyes can see you coming from 50 metres away. For a ghost crab, the horizon is not a mere line in the sky but a full circle. They can be moving forward, strolling on the shore, and instantly change direction and skedaddle sideways when they want to, dashing off at 15 kmph, faster than any land crustacean. They move and live and breed to lunar rhythms—feeling the moon's tug in their blood thickening and thinning, moulting and growing during full- and new-moon phases. Watch them throughout the day, and their 'lunacy'—that is, moonliness—becomes more apparent. They change the colour of their carapace according to background brightness—paler during the day and near full moon, distinctly darker at night and near new moon. They are compulsory company to all of Chennai's beach-goers. They may cross your path, scurry into your slippers, scour the leftovers from food shacks and picnics. But their indispensability to a sandy beach ecosystem remains entirely unsung among the general public.

Beach sand has very little microbial life in it. A washed-up eel or a stargazer fish can lie rotting in the open for weeks if ignored. If not for ghost crabs—the principal beach clean-up squad and public health officers of sandy shores—beaches would be less liveable and hygienic places for numerous life forms, including us. They are crucial scavengers here, the vultures of sandy coastal ecosystems. A 2014 study by biologists Serena Lucrezi and Thomas Schlacher shows that ghost crabs facilitate major transfer of energy from marine ecosystems into sandy and dune habitat food webs, making them a keystone species.[32]

I also like to call them land pirates in the most crabbish, voguish sense. They snatch hatching baby sea turtles crawling to

[32] Serena Lucrezi and Thomas Schlacher, 'The Ecology of Ghost Crabs', *Oceanography and Marine Biology*, Vol. 52, 2014.

the ocean, steal shorebird eggs and chicks, and if times are hard, they may even eat each other. Their burrowing bio-turbates the sand, mixing up and distributing minerals and nutrients, making the subsoil more liveable for other species. Rivers bring sediment. Wind and waves make the beach over millennia. These ten-legged beach-keepers are the beach's indispensable maintenance crew.

Out of the estuary, a grey mullet torpedoes 3 feet above the water, catches the sun's glint, and splashes back in. A viola hermit crab examines the rim of a discarded pill bottle. A fisherman drags out his hand-cast net, which has caught shivering anchovies and mackerel. A small cloud of little stints comes into view over the ocean, 'ssshhing' softly overhead as they swerve down into the creek. Down on earth, tower snails have screwed themselves upside down into the tidal flats, their black opercula (lid) facing the sky as they wait for the tide to rise. Numerous discarded face masks wash up at the river mouth, some ear loops entangled in razor clams, murex shells and dismembered pincers. Four beached sand stars lie in a mysterious square outside a ghost crab burrow. The creature must have been dragging them towards its lair before the sand thudded with giant bipedal footsteps, making it abandon them for the time being and vanish into the ground.

Section 2

TREE MEDITATION

Urur Kuppam Beach/Adyar Estuary
8 April 2021

Along the coastline today I keep seeing ownership strewn like spindrift. In the crab nets brought in there are several red-socked hermit crabs and orange-striped hermit crabs that are using the shells of grey bonnet snails, fringed frog snails and turnip snails. On each of them there are at least three or four hermit crab anemones (*Calliactis polypus*) and Indian measled anemones (*Neoaiptasia commensali*) which have all withdrawn their tentacles once removed from water and look like large skin warts. I find an ark shell on the sand, almost unidentifiable, clothed in furry brown algae and patches of soft coral. Then one half of a pen shell (*Pinna bicolour*) holding on its interior side the hatched eggs of a spiral melongena snail, which looks like the skeleton of a small basilisk from *Harry Potter and the Chamber of Secrets*. The edges between entities and their rights of possession are consistently difficult to trace.

A few of the nets have snared several decorator worms from clayey patches of the ocean bed. I open the heaped tangles on Velan's boat—my friend from Urur Kuppam who moonlights as a fisherman and works in an IT company during the day. Decorator worms are crossdressers, supergluing to their tubular cases what fragments they find and fancy on the seafloor. On one worm's shell-shard-stack attire there is a tiny, brittle star shying into the gaps, and what looks like a porcelain crab having just grown out of its larval stage. Pieces of wedge clams stuck on the worm's clothing are growing acorn barnacles. I find another worm in Velan's boat with a different sense of style. It has used a bit of green plastic from a single-use bag; blue plastic from a tarpaulin sheet; thick, clear packaging from an Amazon parcel; black plastic from a meat bag; a piece of a betel-nut sachet and a Sunsilk shampoo sachet; then some shell fragments, sea-fan bits and nylon strands. It wears what it has found and liked in its corner of the seabed. The fashion sensibilities of decorator worms are changing.

Some boats are returning from the *paaru*, underwater hillocks which lie two hours away from the shore. They bring reef- and rock-living fish like *ora* (rabbit fish), *koli ora* (yellowfin sturgeon), *kal sankara* (blue-striped snapper) and *pulli kalavan* (spotted rock cod). Over the last several days the boats have also been bringing back bycatch rock pieces, gastropods, sponges and other things that appear to have been stroked by some marine Midas. Golden tunicates (*Symplegma brackenhielmi*) are common, especially in summer. They are small breathing tubes living in colonies, glazing themselves over other things, like the scales of a monitor lizard, but gold. They also envelop jetsam, sunk metal and moorings. They are known to fare well and spread in the feverishness of the ocean. They are creatures whose native range is unknown but through ship travel have invaded warm and warming seas. Right now I wonder what those marine hillocks look like. Are they entirely gold-scaled?

At the Ennore Creek in North Chennai, fisherfolk from eight villages worry about another invasive species spread by the ballast water from ships at Kamarajar Port. The charru mussel (*Mytella strigata*) from South America—called *kaka aazhi* here—is taking over the tidal flats in the Kotralayar River's lower course.[1] They are evicting the life-laden *vellai aazhi*—the Indian backwater oysters and the reefs they form, where prawns come to breed from the sea—and the beds of yellow clams. Kumaresan Anna from Ennore describes how the kaka aazhi seem to thrive in the ash-covered bottom of the creek—the ash leaking from nearby thermal power stations.

Tunicates represent a blurry edge between two evolutionary seas, one ancient and one fairly new: invertebrates and vertebrates.

---

[1] *The Hindu*, 'Fishermen Raise Concern over Invasive Species of Mussel in Ennore and Pulicat', 27 December 2022.

There are other physiologically intertidal zones they occupy. They are hermaphrodites; they are also the simplest animals with notochords, that is, primitive spinal cords. Their larvae look like tadpoles. Their very physiology is a deep time junction. Tunicate outer bodies are called tunics, made of cellulose, the plant protein. Their glazings on rock and gastropod shells speak of a warming ocean.

My writer friend Siddharth Pandey did his PhD thesis on magic and craft in literature.[2] He points out how in all magical imaginations across cultures and ages there is one notochordal commonality—everything is alive, everything is speaking. Human agency exists amongst these entities and is often frail in the magic of multispecies world-making. The ents in *The Lord of the Rings*, the alive ocean and winds in Ursula Le Guin's Earthsea series, the terrain itself in Terry Pratchett's Tiffany Aching series, everything from the wand to the willow tree to the dark forest in the Harry Potter books, and so on. But to think of it, if one observes and can deeply listen, the ocean and the living earth are intrinsically a magical place. From hermit crab to rock cod to the marine hillocks to the golden tunicates. Everything lives, everything speaks.

Velachery Marshland
7 May 2021

The summer koel rises before my alarm goes off. By 4.45 a.m. or earlier he is already calling in full breath. His steep crescendos strike the thick twilight air like a gong. I've been waking up bleary-eyed the last few days, often with a headache. He disturbs the sleep of crows too, and they chase him away

[2] Siddharth Pandey, 'Emplacing Tasks of Magic: Hand, Land, and the Generation of Fantasy Taskscape in Terry Pratchett's Tiffany Aching Series', PhD thesis, Cambridge University, submitted 2020.

after he sings for a while. Indian ash trees are beginning to fruit. Their early-morning silhouettes are sharp and eerie. All their leaves have fallen. Eye-shaped lenticular bark nodes stare from all around their bows and trunks.

Sweat bees (*Nomia westwoodi*) first began nesting in my balcony in March 2020 in a tulsi pot, just as we were going into lockdown. They've been with us for over a year. In August 2020 their pot broke while we were transplanting the tulsi. We replaced it, and the bees moved house within weeks. At the time we had five plants. My mother has now made it fifty. I've spotted the bees also nesting under the hibiscus and the borage. Four sister bees share the tulsi pot now, their abdominal bands as blue as hot welder flames. Today they are first active at 6.15 a.m. One sister comes to the mouth of her turret and stands there twitching for several minutes before flying out. Others do the same.

A snub-nosed crow startles me. It catches the grill and gives a long, stern caw, demanding early breakfast. I set out bits of bread which it stabs and eats with its beak-less mouth.

The bees fly from the balcony down into the fallow land opposite the building where I live. But it is difficult to hold sight of them beyond a few metres. Bee flight, bee time—too fast for my humongous human eyeball. A blink is unduly long. They return with the scopae (pollen bags) on their hind legs and bellies turmeric yellow with pollen. They always hover and drift-dance in broad shapes of eights and Zs for a considerable time before dropping down and entering their burrow. Sweat bee sisterhoods don't have a caste system like in other bee families. Females simply share a nest together.[3]

A bee with a hurt wing is crawling all over the balcony floor, trying to fly. Watching it I remember another crippled bee last year which for two days made incessant efforts to fly, before

[3] Arati Kumar-Rao, 'No Honey, No Hives, but Solitary Bees Have Important Lives', Mongabay-India, 8 March 2021.

dying. Bees break their wings at times, flying continually back and forth from burrow to flowerbed. Nectaries of flowers have been found to shrink as the climate warms. Bees have to make more and more nectaring trips as years pass.

Two sunlit pelicans fly towards the Kallu Kuttai Lake to the east, their wingbeats broad and majestic like pterodactyls. Another glides fast behind, catching up, the wind strumming its glowing throat pouch. A whole sky of tall wizard-hat clouds are moving west in unison, making my gut think that the earth is turning faster below my feet.

By 5.40 a.m. every morning an impeccably punctual francolin is awake near my apartment. The bird cracks open the dawn and splits apart the stone-set stillness of a summer morning. Its post is a thick question mark of a prosopis root. Brush clearing on this plot of land has been paused by COVID-19.

I interrupt my workout today to watch from my slide window this bird, flecked like rippled mud, built like a rock. It flaps into a blasting flight like a jackhammer. When the francolin shouts with its neck outstretched you can see its syrinx volley violently up and down its cannon-barrel throat. A comrade from the grass fields across the railway tracks responds almost immediately. Then another yells from a derelict plot two apartment blocks away. Amongst them they seem to create a map of calls, which they pin on and affirm at the beginning of the day—as though to triangulate each other. The bird climbs down the root and sets off within brief minutes.

Next, it is a mongoose on the prowl. I see it today slithering in and out of the tall dead-thorn pile. A mongoose never walks the land but bounces across it on spring-loaded steps. Its form is ever in electric hackles, tail taut, ears always erect. A live wire of a creature, a nerve on the move. It brings a raw wildness even into these concrete squares. I often see it as my spirit animal, my patronus. The mongoose is a hero for my inner child who rebelled and ran away from home, found cracks within the structures closing in upon him and slipped through.

Velachery Marshland
3 June 2021

Moths and pipistrelle bats flicker under the streetlights. A lone thick-knee whistles from across the far darkness.

Pathways of purple light later cleave the dawn sky. It seems like the koel has been singing all night long and into the dawn. The empty railway corridors roll back the kingfisher's calls in three different echo trains. Ribbons of little swifts chase each other and play, chittering brightly. The fruiting Indian ash tree quivers with squirrels. And the sky ripples with massive blades of pelicans, the largest flocks with up to forty birds, folding into crossbows. They flap in slow pulses during their glide. The first bird starts, then the second, the third, and a balletic wave of wingbeats moves across the flock. Like a slow-plucked string, casting a stillness upon the land. The cormorant flocks higher above fly in writhing, chaotic lines of hundreds. I see at least 200 pelicans before sunrise, all steering south towards the Pallikaranai marshland. The arms of their fluid Vs are always much longer to the east. The flocks swim down, over and between stacks of pylon towers in the distance, like vast eels. Sky snakes. Above me the boulder chest of each bird turns blinding white as the sun climbs over the rooftops, further filling the sky with painted storks, black kites and more cormorants. A detonation of pigeons from the roof of the train station, for reasons unknown, makes a stork flock veer away.

The sweat bees are up by 6 a.m. sharp. So are the paper wasps nesting in the vacuum cleaner bag. With my binoculars I track some of their flight paths to the large neem tree diagonally opposite my apartment. Its canopy is about 30 feet away and flowering. The neem holds its small white blossoms on its terminal branches for the paper wasps, plain tiger butterflies and several bee species to navigate and land on easily. The ant- and bee-pollinated cannonball tree lowers its intoxicatingly fragrant flowers down, just a few feet from the ground, on long, hanging stalks for its insect allies to reach. The sausage tree and the barringtonia hold their flowers on hanging rope ladders for little non-winged beings to climb up and down.

Sweat bees are called buzz-pollinators.[4] Their sonication acts as keys to loosen pollen on the anthers of several flowers or move them out of small slits or pores. The insects use their abdomens and flight muscles like sound-beam guns and shake out the pollen. Otherwise it is difficult to dislodge, or not possible at all without the resonating, buzzing bee frequencies. Buzz-pollinated plants include tomatoes, brinjals and several other common vegetables and fruits. Also potato, kiwi, blueberry and cranberry. Almost 9 per cent of wildflowers are known to be buzz-pollinated.[5] These are not easily pollinated by social bees like honeybees which don't vibrate as distinctly as several solitary bees.[6]

The francolin gives a warning call—a short, shrill scream. Squirrels and mynahs scold something. I search the trees and the ground keenly for a rat snake or a mongoose, but it is just the grey apartment cat prowling on the wall of the compound.

By 10 a.m. the sun presses over my shoulders, near 40 degrees celsius. Black kites sky-lift rapidly. They turn into hazy soaring spots within minutes. The Perungudi landfill in the distance scorches into a cloud. The bee sisters are now coming back with legs covered in dark, mustard-yellow pollen and I guess they've been going to the flowering copperpod trees further down the street.

---

[4] Paul A. De Luca and Mario Vallejo-Marin, 'What's the "Buzz" About? The Ecology and Evolutionary Significance of Buzz-pollination', *Current Opinion in Plant Biology*, 2013.

[5] U. Amala and T.M. Shivalingaswamy, 'Role of Native Buzz Pollinator Bees in Enhancing Fruit and Seed Set in Tomatoes under Open Field Conditions', *Journal of Entomology and Zoology Studies*, 2017.

[6] Stephen L. Buchmann, 'Bees Use Vibration to Aid Pollen Collection from Non-Poricidal Flowers', *Journal of Kansas Entomological Society*, 1985.

Velachery Marshland
6 July 2021

I have lunch sitting with the bee sisters at noon today, between back-to-back classes. Virtual pandemic school is taking a toll on my lumbar spine from too much sitting and too many Zoom meetings. As I eat I see for the first time the bees scouting new ground for a nest in a hibiscus pot. Six of them seem to agree upon this spot and are hovering and landing there. They roll over little rocks and clots of mud with team effort, scratch the soil with their forefeet to check its softness and move away dry leaves. One sister gets stuck under a tiny stone which her co-nesters roll over her as she tries to tackle it from the other side. She flies away uninjured when I gently lift it off. Sweat bees never stay on the ground for more than 10–15 seconds. As a habit, almost as a precaution, they take off and do a buzz dance or sortie around the pots and then land again to resume work.

I presume that their previous nest—one with three turrets under the tulsi plant—has sent off several more of these bees, these tiny blue blessings, out into the world. Maybe their daughters are among this crew.

A tent web spider (*Cyrtophora sp.*) had built her holographic web beneath the hibiscus several days ago. She rests like a dark green tick mark at the centre of its architecture. The web is littered with mosquitoes pale from the rains. In another pot nearby my mother has planted butterfly pea seeds, but the soil inexplicably bears little clusters of oyster mushrooms.

I have nearly no memory of the sky, having spent hours mostly in front of screens. On the ash tree I see a dull grey hunch-backed shikra like a sitting shadow, watching the broom grass, almost complaining about the sun. Its attention towards the ground is invulnerable to the mynahs and squirrels protesting its presence.

I learn a bit of squirrel-ese every day. I seem to have tuned my ears to their calling so well that I hear them frequently through Zoom calls—a squeaky background metronome. Today a palm squirrel close by a colleague's window causes feedback noise, making others yank off their headphones. At least one squirrel—if not two or three or five—gambols on the Indian ash tree's branches all through the day. Scarf tail to the clouds, eyes almost always watching the ground, clutching upside down. It jumps on the compound wall where neighbours below keep rice for crows. When it eats, it sniffs, nibbles ten times a second. Moves in blurry, fast-forward, stop-gap frenzy. Lives in a flurry. It is the neighbourhood's watch-tower guard, treetop sentry, whistleblower. Edgy, staccato squeals and rapid tail flicks while wide-eyed and earthward-looking indicate for sure that there's a troublemaker around. A cat, a snake, a mongoose, a meddlesome human—often me. As the prowler moves, the squealing accelerates; slows down when the action fades. I keep an ear out for this every day.

At about 2.30 p.m., after a class, I check on the bees again. At one end of the hibiscus pot the coco peat is swept clean and the debris around cleared. A slanting hollow has been dug in the coarse mud into which at least four bees fit. The white hair on their lower backs and their wings have turned brown with dirt from all the digging work. They have not flown out today. Not carried yellow pollen on their bellies or hind legs, because they are just setting up their nest and have no larvae to feed. I watch them at work for a little while longer, until the next meeting.

Mask tan …
the mud on the caps
of mushrooms.

Velachery Marshland
17 July 2021

The first wandering gliders in the sky for the season, I mark in my diary. Tomorrow the Tamil month of Aadi begins. The farmers' festival of Aadi Perukku celebrates the escalating monsoon in the Western Ghats, and a bit of rain for us,[7] which results in the swelling of river basins and lakes in the eastern plains. The dragonflies are here to tell us this, without any doubt.

It has been pouring every evening for the past few days. Ant files scurry out with their long larvae and pupae, shifting nests. Nights are made of termite alates clouding gold around lights, invading any open window. Puddles mark land like fingerprint powder—showing up all the troughs, tracks, tyre marks, creases, raises. Nowhere is ever flat, they reveal. But where are all the frogs? None still, except one large toad in the carpark.

Yesterday I found a bee sister crawling in the hall with a crumpled wing. I left her back under the curry leaf plant where there's a busy nest with three turrets. There are at least five or six nests in the pots on the balcony. One which four sisters worked over for days, digging under a hibiscus, has been abandoned. They've now occupied the next pot—an ornamental plant with dark purple leaves which a neighbour gave us.

Today I was at a nursery buying plants for a butterfly garden. I saw another gang of sweat bees there, along with honeybees, hoverflies, lime butterflies, nectaring on a variety of flowers. I watched them for a while, tried to follow their flight paths, delighted to see another group of sisters kilometres away.

Cluster fig trees in the nursery had fully ripened fruit. The smell of trampled figs was mouthwatering. The floor was birthing wasps and was full of foraging insects. Around the crowns were clouds of wandering gliders and picture wings hovering to announce that evening's rain.

[7] Tamil Nadu gets a bit of rain from the southwest monsoon between July and September, but most of its rain is from the northeast monsoon between October and December.

Velachery Marshland

28 July 2021[8]

With its beak tilted skyward, a magpie robin sings from atop the lightning rod in the next building. It's a snowball storyteller. About half an hour ago, it had just begun—two sweeping notes descending, followed by one rushing up and left hanging, unresolved. But now it has built on this motif and is rendering an elaborate song. I've come for a run on this road, which has fallen silent post-pandemic. I am accompanied by dog walkers and fellow joggers. A coucal sits on the divider and hoots deeply. The neem trees on the roadside murmur in a voice which is theirs alone. And from above, the magpie robin sings on most mornings. Yesterday its song was starkly different, as my phone recording testifies. The theme of its tune rose and fell, alternating, and over time was embellished into a syncopated song. I remember when, as a student, I went birding in the school campus during breaks and thought on many occasions that I was hearing a new bird. But when I excitedly searched for it, it would turn out to be a magpie robin. As I jog closer to home, robin song tapers and I enter cicada zone. Their loud buzzing from the rain tree trunks is on par with the boisterous local boys' cricket match.

The magpie robin is an improviser. It never repeats a set song. It creates its music spontaneously each time it sits to sing. It chooses a couple of notes—hesitates—adds a few more—inserts inflections—plays with rhythm—pauses. Starts again—tweaks its tune—adds a nuance or two—shifts direction—stops. Starts again. Give it enough time and it will create a complex and original song, which it will keep varying and furnishing till its singing schedule is over and it has to go about its foraging or other work.

---

[8] This passage first appeared in *Sanctuary Asia*, Vol. 40, October 2020, as 'The Stories Songbirds Tell'.

There are many improvisers among Indian birds aside from the magpie robin—the orange-headed thrush, sibia, mountain tailorbird, yellow-bellied fantail, shama and Abbott's babbler, to name a few. They sing spontaneously, composing on the spot. The Indian word for improvisation is *alapana*, which is a Pali and Sanskrit word for 'conversation'. Through their voices, birds dissolve the distinction between song and speech, making it difficult for our forms of description. This asks deep philosophical questions about songbird psyches. Neuroscience describes improvising as 'one of the most complex forms of creative behaviour'.[9] If so, birdsong then beckons enquiries into our notions of sentience. Is it something to be measured on a linear scale? Does it run on many planes? Do our own boundaries of sentience limit the definition of it? Birdcall researcher Jill Soha has discovered that several birds learn to sing by improvisation rather than imitation. They learn to sing by mostly figuring it out and playing with their voice on their own, which brings to mind the famous words of the jazz improviser Miles Davis, 'I'll play it first and tell you what it's called later.'

Another haunting and emotive singer among Indian birds is the Malabar whistling thrush of the Western Ghats. Its melody imprints in the mind and carries in the ear for long after. Its flow and fluidity evoke a river meandering in the hills. It chooses its podium—a stump shrouded in the mist near a forest stream or somewhere where the reverb is best in the wooded soundscape. The first time I heard the Malabar whistling thrush was when I strayed off alone from our campsite during a school field trip in Class 9. At the time I didn't know of this bird and wasn't able to spot the source of its clear, lilting call in spite of frantic searching. It felt so human-like that I began to think it was something paranormal and started a brisk walk back to the group.

[9] Roger Beaty, 'The Neuroscience of Musical Improvisation', *Neuroscience and Biobehavioral Reviews*, 2015.

We are an ocular-centric species—especially the city-dwelling kind among us—leading screen- and chair-based lives. We have come to treat vision as the only viable window of perception. Our other sensory portals—hearing, smell, taste, touch—are fading in their use, in the ways only they can invoke the world for us. There is interesting research emerging on how 'ocular centrism' and the superiority of 'vision' in our society have played a role in creating dualisms in our perception—mind and matter, object and subject, intellect and affect, rationality and emotion, etc.—by the very nature of how vision functions, divorced from its sibling senses. But songbirds, whales, cicadas and bush crickets are phono-centric species, just like numerous indigenous communities with oral cultures which have existed for centuries without the written word. Are they then able to experience a less fragmented and divisioned reality? Perhaps. Navajo rights activist Shelly Lowe's statement 'what's missing is not voices, but ears'[10] has a civil rights context; it is also an apt reference to our relationship to and the sensitivity of the more-than-visual world, or the lack of them.

In January 2020 I was on a field trip to Kaziranga National Park. During a late-afternoon congregation of Jeeps in the Burapahar range, at the junction of grassland and marshland, rhinos, elephants and water buffaloes could be seen in the distance, wading deep in slush. Nervous hog deer scooted out of the nose-high grass. Dozens of cameras waited for a tiger to come out of cover. A Jeep driver remarked that the big cat had just eaten and was not going to budge. Throughout this wait, high in the sky above the pachyderms was a tiny drifting, fluttering form singing its heart out. A shrill, fervent and complex song modulated by wind gusts. The oriental skylark for its courtship display flies up

[10] Quoted in 'Indigenous Participatory Design Toolkit', MSU Library, Montana State University.

into the air till it's a dark blemish in the firmament and renders its elaborate piece while incessantly hovering.

I've been scouring through eBird, listening to skylark recordings from different parts of India. Each birdcall is shown as a spectrogram, which is like a sound river flowing across the screen. The positions of black lines on the graph indicate frequency. The higher they are placed the shriller is the pitch and vice versa. The intensity of the lines indicates loudness. And this illustrates visually the shape of a bird's call. The magpie robin's song is a theme-and-variation, which comes back to the same pattern after each strophe and then varies and elongates. The whistling thrush's song follows a smooth, stepwise motion, with the occasional fitful inflection. Skylark's song, however, is an assortment of distinct sonic patterns one after another. On closer listening, it's a progression of different calls—a mimicked repository of local birdcalls. Some of us listened to the skylark during our long (unsuccessful) wait for the tiger. Now you hear a drongo, then an ashy prinia, this sounds like a thick-knee, this is surely a francolin, now a tailorbird. In different geographies this bird's staggering aerial mash-up displays seem to consist of common birdcalls in those regions.

Air, atmosphere, soundscape—these are nearly non-physical to humans. To songbirds and other vocal creatures, these are essential for their communication, courtship, territory-marking, expression. A couple of years ago, an interlink road was constructed opposite the apartment building where I live to decongest traffic on two outer main roads. Across from my balcony is a large sprawl of fenced-off, unbuilt land, dense with neem trees, ash trees, prosopis and grasses. In the mornings, the pied bushchat and the magpie robin would sing from the barbed-wire fencing. On patches of sand and around puddles a pair of pied wagtails would waddle about and serenade brightly. The road came up adjacent to this habitat. Roads, vehicles and machines treat soundscapes as dumpsites. With the incessant engine rumbles and blaring horns, the birds vanished quickly, even though their physical habitat remained intact. Noise denied them their right to sing.

Emerging studies reveal how birds are changing their songs, pitching them higher and louder in city soundscapes, to be heard. Other reports show that songbirds simply vanish, move away beyond certain thresholds of anthropogenic noise.[11] Lockdown, and the brief quiet it has brought, seems to have invited the magpie robin back to my street.

'With vision we are at the edge of things. With sound we are always in them.'[12] The perceptual fields of numerous birds are profoundly auditory. Most of their experiential world is 'sound' in its plenteous shapes, forms and meanings. Theirs is a world-hearing much more than a worldview. And by paying attention we may perhaps discern the edges of this, nurture an empathy and kinship for beings with other ways of knowing and feeling. What vividness does the world acquire through a thrush's ears? What inspires a magpie robin's new tune each morning? What might a warbler feel when a highway appears by its wintering grove of trees?

To listen is to wet one's feet on a different shore of perception.

Tree Meditation
Kotturpuram Urban Forest

Three members of the public turned up for a poetry session I organized at the Kotturpuram Urban Forest. One left in the middle. I sat under the crepe myrtle trees asking for guidance and wrote a tree meditation in their shade which I used to begin the

---

[11] Victoria Gill, 'Traffic Noise Impairs Songbirds' Abilities', BBC News, 3 February 2021.

[12] Tim Ingold, *Perception of the Environment* (Routledge, 2000).

workshop. Later, I sat under the large rain tree in the middle of the forest and read it out to the two who had remained.

(Tree meditation to be read aloud in the presence of trees, with pauses between sentences and paragraphs, leaving enough time for participants to visualize. At the end of the meditation each person can share a line or two about the tree/s which arose in their mind, and these can be stacked to form a collective lyric.)

Gently close your eyes. Take a few deep, conscious breaths.

As you breathe, visualize how air enters your nose, branches into your lungs, enters your blood, then travels and gives life to each of your cells.

Feel each cell in your body drawing in the air and giving it out …

As you breathe, visualize this breath emerging from the trees around you, from the small shrubs and the many grasses. Visualize breath emerging from every leaf from the canopies that surround you—entering the air, entering your lungs, then your cells—letting you be alive and conscious. Imagine this process as you breathe. Breath from the trees also entering the butterfly, the woodpecker, the earthworm, the mongoose and all the other creatures living in this habitat … breathing the same breath as us.

And as you breathe out imagine the trees breathing in. As you breathe in, imagine the trees breathing out. As you breathe out, the trees breathe in, making your breath into wood, into leaves, into roots and bark. Bring to mind the creatures living in the trees—eating leaves, nesting on branches, hiding in bark and hollows, resting in the shade. All of them living and breathing together. Visualize this shared air, shared breath with all other species. Visualize how the breath and the lives of all these creatures, including us, converge on the trees. Imagine this as you breathe.

Now enter your mind and see how trees branch into your memories, your emotions. Is there a tree which holds a special place? Which is the first tree you spent time with? Is there one of which you have strong memories? How did that tree touch your life and how did your life touch the tree's? Where did it stand? What was its shape? What life did it hold? Bring that tree vividly to mind now as you breathe ... Invoke the feeling of that tree as you breathe ... its form, its life, its energy.

Kovalam / Muttukadu Creek
13 August 2021

The pre-monsoon rains over the past month have begun flushing out all the riverbeds and creeks along the Coromandel Coast. Sediment from them has created sandbars which have blocked all the estuaries and river mouths in north Tamil Nadu. If one is to take a walk right now along Chennai's coastline, one can do so without having to cross water.

Four naturalist friends and I are at the Kovalam Creek, which is about 30 kilometres from Chennai. Where a gushing stream used to flow from the creek—which is contiguous with the vast Kelambakkam backwaters and salt lake—there is a sand bridge several metres wide. We cross over to the other side. Ladyfish are stuck in pools, squirming and slapping sideways over sand to reach the incoming tides. Fisherfolk cast their nets into the blocked channels and the waves. They say that an excavator is needed to dig out the creek mouth, if not all the fish in the backwaters would die. The intertidal gleams with rich black magnetite, making the sand hotter to walk on in bare feet. Rohith and I have a running race till the next groyne.

A carpenter bee truly keeps us company and flies with us for a long stretch as we beach-comb. Carpenter bees are common on the coast because driftwood creates a habitat for them. Their sisterhoods build hollows in them and get pollen from beach vegetation like goatfoot vine and sea beans. The haze of salt and sea vapour, pluming up from crashing water like yarn from a charkha, makes hair wild and skin sticky. Greater and lesser crested terns fly inland in loose, lazy troops. Common terns perched on thermocol flotsam bobbing on the waves seem like porpoising dolphins in the distance. I shout 'dolphin' once, only to find a black buoy being tossed by the swells with a perched common tern. A soaring osprey dives with stretched talons down at the backwater, where some sixty painted storks stand idle. Then it breaks off halfway and flies into the casuarina trees.

The ocean bed is distinctly different at Kovalam, more rocky and coralline than in Chennai. Amidst the strandlines and shells we find cattle ribs, the volcano art of barnacles, sea turtle beaks, beached coral and sea-smoothened bits of coal that have drifted from power stations or container ships. The beach-loving Bibron's skink jumps out of a crevice in the groyne and shows its candy-orange flanks, takes in a bit of sun and swims into the granite again. There is also a ghost net full of entangled murex snails and a cloud of flies around their stinking, dying bodies. Murexes prey on bivalves and other sea snails on sandy ocean beds, and their elaborately spined shells deter predators. But precisely what makes them invincible in the ocean is their Achilles heel when met with the human world. More than any other creature, murex snails are bycatch in nets and are snared in heavy dozens by ghost nets and discarded clothes drifting in the seabed.

I spot the kills of a familiar pair of pied kingfishers on a row of large laterite rocks found on the mudflats at the cheek of the creek. Empty skulled mullet which has been clobbered on stone with sharp rock oyster casings. While crows quarrel over these leftovers, we have our packed breakfast here. Between these stones, among other things, are the wondrously polymorphic dubious nerite snails (*Clithon oualaniensis*) whose shells come in as many patterns as shirts in a clothing store.

Out near the fishing hamlets of Kovalam the last boats are arriving. We see a lot of bycaught hermit crabs carrying their cherry-red egg sacs under their abdomens. It is their breeding season. My sighting for the day is the sponge crab (*Laurodromia sp.*), furry as a drenched rodent and with nail-polish-pink pincers. A crustacean with deep black eyes, it can host on its carapace algae, soft corals, sponges, sea squirts, anemones and other creatures. The toothpick crab (*Phalangipus sp.*), which is the daddy longlegs of the crustacean world, the olivid hermit crab (*Diogenes miles*), which uses only the shells of olive sea snails, and the marbled crab (*Neoxanthus michelae*) with its honeycomb carapace that we see in the nets are typical of the rock beds in this part of the sea. Some of the rock is natural to the seabed, some are boulders storm-rolled from the groynes.

Velachery Marshland
15 August 2021

On the empty plot by the train station, candle bushes (*Senna alata*) are aflame. I gaze at the dense understorey of their incandescent liquid-yellow blossoms from my terrace in the afternoons when I go to get my dose of panoramic vision. I've been visiting the candle bushes for the past few days to watch all the insects which come to their glow. Their massive compound leaves fold up like books at dusk. Emigrant butterflies are busy right now laying eggs on the fresh leaflets.

The male emigrants fly with a jerky, impulsive attitude while they gallivant for females. Sometimes I come across a male left in a daze when a female flies by while the male is pursuing another, unable to decide which one to follow. And it is not uncommon to see an airborne party of three or four males in reckless, undulating pursuit of a single female. In such a chase, the males equally invest their energies in keeping up with the female, and brawling and bashing one another. For a female it can be a mighty struggle to lay even a single egg. Before she can locate a leaf and curl her abdomen forward, one or more desperate males will be on her tail, who she must now fly around the bush and lose before she can think again of oviposition. If it comes to it, the female will fall to the ground, hide in the vegetation and feign death.

This patch of land, about 100x40 metres large and partially in the shade of the railway station's rafters, has grids of 4-foot concrete columns where further construction has been abandoned. The half-done underground foundations have become square ponds where duckweed spreads and typha reeds grow. They are full of granite ghost dragonflies—creatures which vanish from vision as soon as they perch on cement or rock. There are leucas plants (*thumbai* in Tamil) flowering. Three species of carpenter bees, banded bees, cuckoo bees, stingless bees, at least two species of honeybees—the creatures seeding, feeding the world—are all convening on the flowers of plants growing on land left to itself and open to sky.

Soon, sweat bees arrive and thrill me. Dark eyes, reddish legs, bulging black thoracic plates like backpacks. Banded bees behave like hummingbirds—dancing, hovering too much, swerving sharply, sitting at each flower for milliseconds. Cuckoo bees are even faster. They are finished with five or ten flowers and gone before I can get to them and crouch down for a photograph. But sweat bees are slow bees. They fly in straight lines, are averse to sharp turns and acrobatics near flowers. They sit at flowers and crawl patiently into the corolla. They take their time. Patrolling black ants disturb them often. After nectaring for a while, they may even perch on a leaf, either still or stretching out their wings, combing their antennae for several minutes. Sometimes they turn around and watch you.

Months ago when we mixed the soil for our balcony pots, we added quite a bit of coco peat to them. Now, after the rains and daily watering, the mud is yielding delightful clusters of flowerpot parasols—aptly named mushrooms. Their caps, club-shaped at first, turn over the fibre-matted soil like trapdoors and flare out. On Friday I thought I'd create a dedicated mushroom bed—which I will keep wet throughout—in a long, rectangular plastic pot where nothing is growing. I left two-thirds of the soil for the bees and chose a corner where there seemed to be no burrows. I got all the coconut fibre from the kitchen, some cow dung and leaf litter and started mixing it with the top few inches of soil. To my grief there was one nest still there which had two larvae—bee-shaped but white like litchi flesh, legs waving—and one pupa in a brownish casing. I put them back in the soil, knowing that they might not survive. Then little tunnels in the mud bled frantic odour ants whose nest too my fingers had crumbled. Tiny, white, mite-like creatures were spilling out of the soil. And a couple of woodlice, which rolled up like tiny armadillos. What a bewilderingly alive thing just plain soil is. Better left to itself. Today at noon I sowed cosmos, chrysanthemums and sunflowers for the bees.

Velachery Marshland
21 August 2021

4.30 a.m. There are streaks of lightning behind the fog clouds, zigzagging like a doctor's scribbles. On the wall over the television two geckos fight, shrieking like sandpaper. Dawn is dark grey and heavy today.

I see bright barium-green flashing around the thumbai plants. At bullet speed small carpenter bees thrust into the tongue-like white blossoms. Flower—one, two, three, gone. Another—one, two, three and gone. A blink is unduly long in bee time. But there is also the slow, dangling hover of the tiny stingless bees—almost like mosquitoes (called *kosu theni* in Tamil, meaning mosquito bee), their hind legs always hanging loose in a long S.

Thumbai holds its flowers in whorls around its stem in two or three levels. Each flower has two petals—one furry upper bulb and one wide lower petal, which is its insect ramp. Little bees like the small carpenter and the stingless go straight into the corolla. Banded bees and cuckoo bees perch on the ramp and probe inside. Butterflies hang on the ramp and unfurl their proboscis. But sweat bees alone perch upside down, part the petals, tear the ramp and reach the nectaries. I see them often rubbing their abdomens lustfully on the upper bulb, looking like they are fake-mating and ripping up the flowers. I shared my observations with Dibyajyoti Ghosh, a bee scientist working with the Zoological Survey of India, and he told me that sweat bees are two things. One, they are flower mutilators. They rip up the petals of the flowers they visit so that they become inaccessible, unperchable or unattractive to other bees and insects.[13] Two, they can also be nectar robbers in the case of some flowers. Especially with leucas flowers since they perch upside down and draw nectar, they don't actually enter the flower and come in contact with the flower's reproductive structures. So they end up not pollinating the flowers but pilfering nectar which the plant holds only on the condition of mutual benefit.

The sweat bee is a system hacker. See the leucas flowers near my home and you know that barely a flower exists that has not been visited by these bees. With other flowers, though, like cassia and copperpod, sweaties properly pollinate them as they have to enter the flower given their structure. And they also gather pollen to take to their larvae in their burrows.

All the thumbai plants have hordes of ants on them. Honey ants, silky shield ants, carpenter ants, odour ants, among others. Ants don't visit the flowers but congregate at the empty calyxes after the flowers have fallen from all the bee activity—I think for the nectar residues left behind. I've never

[13] Sangeetha Varma et al., 'Nectar Robbers Deter Legitimate Pollinators by Mutilating Flowers', *Oikos*, Vol. 129, No. 6, June 2020.

seen them entering these flowers, unlike the candle bushes and hairy sennas nearby.

Under the flower whorls and the leaves near them lurk lynx spiders and crab spiders. The long-jawed orb-weaver stretches itself thin into a midrib—a line on a leaf—then flares into a full spider when the leaf shakes with a landing insect.

Tadpole Tank
25 August 2021

Fruit flies can feel lonely and lose sleep over not being with their friends.[14] Their physical health deteriorates without social connection, showed a paper in the journal *Nature*. The peculiar sentience and not-too-unfamiliar camaraderie were apparently among the floating black specks—called drosophila—over the guavas and bananas on the table. How I see their flying columns now has completely shifted.

There are schools of ornamented pygmy frog tadpoles down the street. They live in an abandoned cement cattle-feed tank caving into itself, one-fourth full with rainwater. Their spinal cords and eyes glow bronze. Tadpoles from the same batch of eggs mostly stay together. They swim to the water's surface and gulp continuously like fish, feeding on algae, detritus and other floating things which my eyes are too large to notice. If there is movement above, they scatter. Not necessarily dispersing or hiding but just breaking form, jumbling their orientations and depths. They randomize into drifting shadows. Then move up, collect again, align all heads and tails like iron filings. Something thinking, living, stretches across all their seemingly separate jelly forms.

[14] Alexandra Le Bras, 'Effect of Chronic Social Isolation on Flies', *Nature*, September 2021.

Intelligence can extend across larger forms, beyond close-fitting skulls, beyond bodies and entities. A star cluster may have its own larger intelligence. And a flock of birds has a complex intelligence, unconfined by feathers, flesh and space. How then does one explain starlings and the shapes of their murmurations? Hundreds of birds spiralling and snaking in the sky, a cloud of black masses clustering, stretching, folding and evolving in abstract ways. How then does one explain a column of fruit flies, a school of tadpoles? This is given different names—swarm intelligence, eusociality and so on. But can science and its methods ever come to understand this fully? Can the mind ever grasp and confine these phenomena within frames? I remembered Gurdjieff: 'It is possible to see and feel the unity of everything. Attempts to connect these phenomena into some sort of system in a scientific or philosophical way lead to nothing because man cannot reconstruct the idea of the whole starting from separate facts.'[15]

I think about our basic needs as we are taught—food, clothing, shelter. So utterly incomplete. Social connection is a fundamental need for many living beings. Our personal well-being and worth are so profoundly hitched to our relationships with others that one of the leading causes for taking our own lives is social pain and isolation. It can be far more deadly than physical pain. We are like pygmy frog tadpoles and fruit flies. Israeli adventurer Yossi Ghinsberg was stranded in the Amazon for three weeks and had to survive by himself. Asked later what his greatest hardship was, he said it was not the fear of wild animals, insect bites, unforgiving weather, lack of food or a flood in which he almost drowned, but loneliness. He created

[15] P.D. Ouspensky, *Fourth Way: Teachings of G.I. Gurdjieff* (Random House, 1971).

imaginary friends,[16] like the hero in the movie *Cast Away*, to keep himself company.

A granite ghost dragonfly's nymph crawls up a chunk of concrete underwater, darts and snatches a small tadpole within reach. It holds it in its mouth. Another nymph charges at it and it goes scurrying between two rocks. The evidence of the hustle is faintly visible on the algal turf. Tadpoles scatter. Move into more open water. Plumes of sediment fume up and diffuse. Death feels as alive as living. A branching, a scattering of lifeways.

The other highly social creature I share this place with are little swifts. They are pure localites. You can see them all year round over the railway station and these few apartment buildings. They fly in air ribbons, twisting streamers, as they chase each other, weaving in and out of the empty towers of the station. Their jubilant, ever-enthusiastic chittering and playfulness lifts me up at once, letting me fly with them.

I didn't know where they lived. Just looked at them as entities ever of the sky. Last week Rohith and I found their hideout nesting place on the tallest, spookiest roof of the railway tower. Breathing was difficult as we walked up into what was easily decades' worth of pigeon droppings on the walls and floors turning slowly into sedimentary rock. A different kind of dropping concentrated at the centre of the floor pointed to the nests above. Wedged to the corners of the truss beams radiating from the centre, the nests were primarily made of pigeon feathers, wood shavings and mud, among other things, somehow stuck to the ceiling. Six or seven nests were pressed side by side, just as swifts and swallows sit pressed shoulder to shoulder. They streamed into the tower to feed their chicks. As the emptiness echoed eerily with their chirping, they flew

[16] Yossi Ghinsberg, *Lost in the Jungle* (Skyhorse Publishing, 2009).

back out to chase insects again. But for the most part, they seemed to fly just for the thrill of the sky and the joy of each other's company.

Chennai
27 August 2021[17]

In the last two weeks of August, perhaps for the first time in Chennai's history, life science students from two women's colleges (Stella Maris and Women's Christian College) conducted Urban Wilderness Walks (UWW) in twenty-five different localities across the city. From Avadi to Integral Coach Factory (ICF), Perambur to Pallikaranai, Triplicane to Thiruvanmiyur, the public in these places, young and old, were guided into experiencing urban spaces through the lens of ecology and biodiversity.

UWW is an internship I began in July through the MNS for colleges in Chennai, with the help of Vijay Kumar, the secretary of MNS, and Professor Kalpana Jayaraman of Stella Maris. The first batch of twenty-seven interns is almost at the end of the programme. The dream of this internship is to create a citywide network of young naturalists, communicators and resource people, and see if this might in some way shift the city's culture towards that of deep eco-literacy and belonging; to get the public enmeshed in the care for this unique landscape and bioregion. As part of the course, students document place histories by engaging with people in their localities; they survey and map trees, document biodiversity using citizen science portals, participate in field sessions in different habitats of Chennai, discover nature-based learning pedagogies, create

[17] This passage was originally conceived for the *Chennai Photo Biennale* journal, Ed. III, and published under the title 'Urban Wilderness Walks'.

their own nature-education material, and of course, create experiences and walks for the public in their localities.[18]

After the first set of walks, I met all the interns to listen to their experiences and wrote down some of what they shared. Pavithra conducted her walk in Lloyd's Colony in Royapettah: 'I'm quite amazed by the interest shown by all the participants. They were bombarding me with questions … We saw kingfishers, house sparrows, chalky percher dragonflies, touch-me-not plants and butterflies like the bluebottle, common Mormon and emigrant. They were overwhelmed because it was their first time seeing all these.' Yamini had organized a walk for the government school children of Lakshmipuram and found them keenly interested in discovering new species in their neighbourhood. They seemed to have almost a natural curiosity towards the biodiversity around them but did not have the opportunities to further these interests. Keerthika did her walk at ICF. She enjoyed the parts which were not actually planned for but happened spontaneously. Like recognizing a mushroom, seeing an unknown insect and using a guide to identify it.

Robert Macfarlane, a legend of nature writing, describes city life thus: 'Anyone who lives in a city will know the feeling of having been there too long. The gorge-vision that the streets imprint on us, the sense of blockage, the longing for surfaces other than glass, brick, concrete and tarmac.'[19] The vision of UWW is to find gaps within the urban itself, to venture beyond its blockages. And to ask what are the most relevant stories to tell people and children about this city, and come up with cultural retellings emplaced in the living world.

'Walk' is a formative word in UWW. How we choose to move has the potential to shape the world around us. When

[18] Prince Frederick, 'MNS Comes Up with Internship on Nature', *The Hindu*, 30 October 2021.

[19] Robert Macfarlane, *Wild Places* (Granta Books, 2007).

I go to Urur Kuppam Beach, I go up till the shore in the mornings, watch what the fisherfolk have brought in, ask them how the seas are, how the winds are, then walk north till the Adyar River's estuary to see its life and flow. The residents and the Chennai City Corporation have gotten used to car-free Sundays on the beach's promenade. If you visit on Sunday mornings between 6 and 9 a.m., the road is used by dog walkers, joggers, Zumba dancers, skaters, placard-holding campaigners, balloon sellers, yoga doers, frisbee and badminton players, tender-coconut sellers. This promenade on a Sunday morning is a beautiful example of a tiny 'open city', as theorized by sociologist and planner Richard Sennett.[20] The density of people is high and diverse. The street transforms into a social space—because it ensures slow movement. There are many kinds of social mixing and face-to-face interaction between people who otherwise might never meet in 'class'ified urban society. Class confinements are visibly shaken up. There is talking, laughing, arguing, debating, gossiping and making. A richness of life which dissipates once the vehicle barricades are removed after 9 a.m.

I asked the interns about how a walking culture, like on Elliot's, might impact society. Some of the responses I got were—improved mental and physical health, lesser depression and loneliness, lesser carbon emissions, lesser roadkills and more thoughtful resource consumption. Elliot's promenade is a small example of a 'walking culture' or a culture of slow movement. In such a place, neighbourhoods are designed for people, not vehicles, in contrast to how modern urban spaces are planned.

Urur Kuppam Beach helps me imagine how and if walking and other forms of non-motorized slow movement could be a predominant social behaviour. How might that influence city

[20] Richard Sennett, *In the Post-Urban World* (Routledge, 2017).

planning? There would be more trees for shade, more parks and benches. Would cities then be socially and ecologically more inclusive spaces? Yes, I think so. There would be cleaner public toilets at more frequent intervals. More small and diverse kinds of shops and economies would thrive, rather than a few massive mega malls. In places like Kullu and Amritsar one gets a glimpse of what this might be like, where several of their roads are permanently barricaded to cars. Of her country, urbanist Jane Jacobs says, 'Not TV or illegal drugs but the automobile has been the chief destroyer of American communities.'[21]

In places of slow movement, we would know the names of more of our neighbours. Public spaces would be spaces of creation. More leaf litter would fall on the ground. Grass and brush would grow more densely on the waysides—bringing bees, butterflies, sunbirds and skinks into our daily speech and imagination. Trees would live longer. Frogs will be heard. It is at the pace of walking that our body immerses itself in the many levels of connection to the living world. Human interaction has evolved to happen on the horizontal plane. Our experiences occur primarily on the x-axis. Which throws a question to the other strange fallacy of urban planning—verticalization. Stacking us on top of each other has the effect of increasing density while reducing relatedness and relationships.

I watched Siddharth Agarwal's extraordinary documentary earlier this year, *Moving Upstream: Ganga*. Siddharth walked 3,000 kilometres between June 2016 and April 2017, starting from Ganga Sagar in West Bengal and finishing at Gangotri in Uttarakhand. As he journeyed, he interacted and recorded his conversations with the riparian communities. He stayed in people's riverside huts and documented the challenges they face due to 'development'—which, on a river, means building barrages, bridges, canals for larger vessels, and river-linking projects. His film made me think

[21] Jane Jacobs, *Dark Age Ahead* (Random House, 2004).

about how many campaigns demanding and possibly achieving sociopolitical change happen on foot. Siddharth often says walking 'disarms' the walker. I know that my own feeling of belonging to Chennai and deciding to put roots here came from walking through its landscapes and streetscapes.

I think an active citizenry is always, or at least mostly, pedestrian.

Last year a Class 8 child from Abacus Montessori School did a citizen science project. She wanted to find out how many of Chennai's common urban trees could be identified by children of her age. In the group of about fifty children, most could identify only four to five species.

Imagine if, by the age of ten or twelve, each child could recognize a hundred plants and trees of their city. Imagine how that would change the culture and politics of urban living. This is not a large number. Psychologist Allen Kanner's studies show that an average three-year-old American child can recognize a 100 brands, and almost 300–400 brands by the time they are ten years old.[22] These numbers may be a bit lesser in India but are probably comparable.

Amnesia about trees is ironic in a region like Tamil Nadu. It is difficult to navigate 10 kilometres on its map without encountering a place named after a tree or a plant. Take the names of localities in Chennai, for instance: Alandur (*alam*: banyan), Veppery (*vepam*: neem), Perambur (*perambu*: cane), Panaiyur (*panai*: palmyra), Purasaiwakkam (*purasu*: palash), Teynampet (*thennam* + *pettai*; *thennai*: coconut) and so on and on.

---

[22] Allen Kanner, *Globalization and the Commercialization of Childhood* (Tikkun, 2005).

Even a casual study of Tamil place names shows how trees and local vegetation are deeply rooted in people's collective imagination across this wide landscape. I posted this on Instagram, and my comments were filled with names of similar places from across the world that have names inspired by trees. Bengaluru, somebody said, is named after the *benga* tree: *Pterocarpus marsupium*. Palakkad in Kerala from the *paala* tree: *Alstonia scholaris*. Pranay, a friend from Telangana, told me that his native village is Vasalamarri, *vasala* being beams of wood and *marri* being banyan. A person from Maharashtra began listing names of villages from his state: Pimpalgaon (sacred fig), Vadgaon (banyan), Ambegaon (mango), Bordara (bor or ber), Palasdari (palash valley), Umbre (fig) and so on.

Similarly, it is difficult to move in any direction on the map of Tamil Nadu—or maybe, as the trees example brought out, any map—without crossing places named after waterbodies. If you are from Tamil Nadu, think of all the place names which have the suffixes -eri, -thangal, -kulam, -odai and so on.

The UWW initiative hopes to bring back into Chennai's culture its ecological histories. It takes inspiration from Nizhal's[23] tree walks, Jane Jacobs, Richard Sennett and Anne Hidalgo. From the wildflower safaris writer Lucy Jones takes her young daughter on, on her sidewalks, and from the lake walks conducted by Arun Krishnamurthy from Environment Foundation India. From the toxic tours conducted by my friend Nityanand Jayaraman. It takes inspiration from Robert Macfarlane's *Old Ways* and Siddharth Agarwal's *Moving Upstream* and Maria Faciolince's collective cartography project. From Sandip Patil's vision and work for pollinator-friendly urban streets and Marine Life of Mumbai's shore walks. And several other such initiatives on foot.

---

[23] https://www.nizhaltn.org.

Velachery Marshland
28 August 2021

4.30 a.m. Is that just one cricket calling? Not likely. I can hear them everywhere. How do they seem like a single voice and a chorus at the same time? Close your eyes, listen too keenly and the whole cricketscape converges into one song. Mind your business, let their stridulation be the backdrop—then you hear the separate voices, the blemishes, the choristers who don't keep tempo. Crickets in the early morning present an observer's paradox. Attend too keenly and the nature of things changes. Let it be at the edge of perception and they shift shape into something else.

The morning is suddenly cold and the sky clear, though clouds from the east are moving in and closing over. In my dream I see huge wedges of great white pelicans passing over my balcony, their beaks and pouches exaggerated pink. Adults and fledglings. I try to take pictures but my camera turns to black clay, buttons roll, keep slipping from my hand. I have seen a great white pelican only once in Chennai on New Year's Day, 2018. It had lost its way and landed up in Perumbakkam/Sholinganallur marsh. It foraged with the spot-billed pelicans and cormorants, then trailed off and swam alone.

Eleven glossy ibises cross over the balcony close to 6 a.m. They fly in an indeterminate, amorphous flock. A slanting blade, then a deep U, a loose W, then passing above like a sidewinder writhing and thrusting forward without a grip on concrete flooring. Black kites find it difficult to fly in the cold, stagnant air. They flap frantically to keep afloat and fly very low.

The sweat bees find it cold too. At 6.30 one sister comes out, hovers around and goes back into her burrow, deciding to stay in a little longer. The first bee flies out through the grills only by 7.20, an hour later than usual. The cosmos and sunflowers I had sown for the bees have begun to sprout in their pots. I put some crushed eggshells around them. A blue-banded bee, with large green eyes and fat legs, has started nesting in the hibiscus pot. Under a mud and coco peat slab she has made her cave. She is stouter than the sweaties and she flies in clear, straight lines (near her nest, not near flowers). She doesn't dizzyingly zigzag like the sweaty tenants in the nearby pots.

I did my diaphragmatic breathing practice with a humpback whale's song playing in the background, which took me into an indescribable stillness. I remember reading that there are three times the number of spindle neurons—the cells of language—in whale brains than in human brains, and they have had them for twice as long.[24] Whales speak and sing into the ocean, to each other and to all species. We don't know what. Maybe we never will.

There are many states of consciousness, described in different ways and words in different cultures. The waking state (measured as 12 Hz and above on an electroencephalogram) modern neuroscience describes is 'beta', where the brain is fast, normal, reactive, purposive. Beta speeds up and is most dominant when we are stressed, alert, anxious. Then there is Alpha (8–12 Hz), a state of relaxation, creativity, play and well-being. Delta (0.1–3.5 Hz) occurs in deep sleep, where the physical awareness of the world is cut off and the deep unconscious is at work. Theta (4–8 Hz), which is the intertidal zone between wakefulness and sleep, is especially active during prayer, daydreaming, meditation and states of 'spiritual awareness'.[25]

Through my meditation practices set in the natural world I have found how attuned our many states of consciousness are to sounds and ecological stimuli. Whale song, for instance,

[24] Andy Coghlan, 'Whales Boast the Brain Cells that "Make Us Human"', *New Scientist*, 27 November 2006.
[25] Ned Herrmann, 'What Is the Function of the Various Brainwaves?' *Scientific American*, 22 December 1997.

is well known to bring the mind to Theta/Alpha and put it in a state of deep insight and well-being. How is it able to do so? What has happened in our collective evolutionary past that we are immediately soothed by the voices of these marine giants? Similarly, the sound of waves, the dawn chorus of birds, the calls of crickets and frogs, the dense rustle of leaves and branches, the soft, whistling wind between trees or rocks, the flow of rivers, and so many other things. The pelagic depths of the mind recognizes something in these soundscapes and shifts its feeling tone.

At 8 a.m. sharp, cricket song falls silent. Reed warbler, coucal and tailorbird take over.

Kaliveli Estuary
Edaiyanthittu
30 August 2021[26]

Strand lines are made of sea urchin tests. Hundreds of these dried or drying exoskeletons—also called sea biscuits or sand dollars—lie along the Edaiyanthittu sandbar. Fisherfolk say that the *olni vellam*—a thermohaline circulation (upwelling) which brings colder water from deeper parts of the Bay of Bengal towards the east coast—pushes these creatures and casts them ashore. Today their nets are full of six-armed brittle stars. This is the season for smaller fish—silver biddies mostly, but also anchovies, sardines, ribbonfish and tonguesoles. By one of the boats I find a rough carrier shell (*Xenophora corrugata*)—a curious marine seasnail which likes to cement stones, bones, bivalves, small gastropod shells and sometimes bits of brick or concrete among other oddities over its home. I hold this one and feel a stab of grief before I can tell why. I remember my sister. The rough

---

[26] This passage first appeared in *Sanctuary Asia*, Vol. 41, No. 10, October 2021, as 'The Forgotten Reefs'.

carrier has bits of clay urns on it—urns used to carry ashes from cremation grounds to dissolve in the sea.

Other than the two longshore currents—one flowing south to north for most of the year (*thendi*), and another north to south during the northeast monsoon (*vanni*)—fisherfolk differentiate between two other kinds of currents, which reflects their embodied knowledge of the ocean. Olni and *memeri*, the latter of which Palayam Anna has explained to me multiple times but I cannot fully grasp sitting as I do on land. He tells me to go fishing, that diagrams on paper cannot replace the experience of the ocean. In the Tamil month of Purattasi (17 September to 17 October), during the northeast monsoon's peak and cyclonic weather, memeri circulation happens (though it can happen at other times as well; like I said I don't understand this) which is in the opposite direction to olni. It makes the ocean rougher and more dangerous.

Babu, a fifty-year-old fisherman at Edaiyanthittu, tells us that olni pushes up sea urchins, jellyfish and other free-floating creatures living in the depths and casts them on the shore en masse. Babu mentions that he has enrolled his son in a marine biology course at a well-known college, but he feels he learns nothing from the virtual classes going on right now due to COVID-19. 'He might as well come and learn directly from the sea.'

Rohith and I walk along the beach and wade through the creek and backwaters in search of oyster reefs. Rohith carries with him a pansy shell sea urchin's test, a stunning exoskeleton, which looks just like a fossilized flower. Ashy woodswallows chatter petulantly, perched atop casuarina trees or fanning over us with their bat-like wings. Dozens of bee-eaters are trilling brightly all over the drizzling grey sky. We first watch the fisherfolk come in with their boats near dawn and unravel their nets. There are more and more sea urchin tests. A live sea urchin is locally called *parattai* or sometimes *mulleli* (translating to 'thorny mouse', which is also the name for hedgehog).

Under the creek, a maze unfolds. A jagged, craggy, amorphous architecture cloaked in slush stretches underwater. Its sharp edges can slice toes. Contorted petals of rock; a frozen subtidal flowerbed. A school of tiny, translucent fish—each the size of a paddy grain—vanish into the asymmetry of the oysters. Their prismatic bodies refract the early sun's rays in pink, red and blue.

This wetland has among the largest stretches of oyster reefs in northern Tamil Nadu. Rock oysters (*Saccostrea sp.*) attach and grow on the broken masses of Mughal masonry from the Alamparai Fort, fallen palm trunks and driftwood. Indian backwater oysters (*Crassostrea madrasensis*) bloom on mudflats, sandbars and beneath the backwaters. Two mechanized fishing harbours are proposed here.[27] We take pictures of these reefs for documentation.[28]

In old Tamil literature, oysters were called *kizhinjal*, though now the word is used broadly for any shell. *Kizhi* is to tear. Nets cast carelessly into backwaters can hook on to oysters and get shredded—something fisherfolk lament often.

Months ago, the mouth of the Edaiyanthittu Creek was at least a 100 metres wide, allowing the sea to rage in. But in August–September creeks and estuaries get blocked. The pre-monsoon showers flush silt from all the riverbeds and wetlands, depositing the sediment along the shoreline. Fishers call this *Aadi vandal thanni* (water laden with sediment, in the Tamil month of Aadi). This creek too is solidly blocked. Tractors can now drive across where waves used to break and where the mudflats were once dangerous quicksand stretches. So, we gladly walk over.

[27] I later filed a case (Appeal No 14 of 2022) at the National Green Tribunal, South Zone, and the harbours were cancelled by court order due to their many illegalities and their threats to ecology.

[28] V.M. Nagarajan et al., 'Status of Important Coastal Habitats of North Tamil Nadu: Diversity, Current Threats and Approaches for Conservation', *Regional Studies in Marine Science*, January 2022.

A ghost crab has tried to drag in a boxfish skull too big for its burrow. Now it stands wedged firmly over its home with the crab unable to come out. On the other side of the creek we meet Ranjith, another local fisherman, who takes us in his boat deeper into the backwaters where he said aazhi are found. Ranjith picks up oysters, plunging his arm almost a metre under the creek. He knows how they grow and the sandy bottoms they prefer. He says that the smaller oysters are left behind by the oyster-picker women from neighbouring villages, and they grow fast and all through the year. But domestic sewage entering through the Buckingham Canal and agricultural runoff has reduced their numbers significantly over the years. There has also been considerable conversion of salt marshes to commercial salt pans—which has affected oyster habitats and the livelihoods dependent on them.

Around the reefs, girdled horn snails and blue-striped hermit crabs congregate densely. We count 500–800 of them inside 1 square metre quadrats. Their trails are vivid doodles on the mud. Mud creeper snails and larger hermits make trails in the sand like short snakes and tractor marks respectively. All of them are strictly backwater species. Here we find all three important coastal ecological architects growing together: mangroves, seagrass and oyster reefs. Perhaps there is an interconnectedness amongst them we are yet to discover.

A rock skipper crawls into an oyster's lip and hides, just as I wade behind it and get my first oyster cut deep under my right big toe. Five whimbrels whistle as they fly from shore to shore.

I discovered, from several research papers, that the diversity oyster reefs can host is comparable to corals. But more importantly, each oyster filters about 200 litres of water per day, removing algae, phytoplankton and other suspended particles and nutrients from it—making estuarine environments healthier and liveable for other creatures. Without this important ecosystem function, coastal wetlands would frequently suffocate from algal blooms and eutrophication, the reasons for mass fish mortalities and death of other creatures due to hypoxic conditions. Oyster reefs are biostructures, natural breakwaters which protect shorelines from erosion—and in many countries their restoration is an initiative to protect eroding coasts. In many parts of North America, military bases are installing 'oyster cellars' to arrest erosion due to flooding and storms. These reefs also draw down and fix atmospheric carbon through their filtration and bio-deposition.

To understand oyster habitats better, I wanted to speak with somebody working with oyster reef restoration. The coast of Tamil Nadu, like other parts of India, still has some vast oyster formations, though in steep decline. Through a mutual friend at the University of Hong Kong, I reached out to Dr Jay Minuti, a postdoctoral research scientist at the Marine Futures Laboratory.

Jay's Instagram feed is enviable. Her posts and stories show her exciting fieldwork. She is either walking and surveying vast oyster beds on the mudflats of Yung Shue, Hong Kong, or is on a ferry mid-ocean, if she isn't measuring sea urchin parameters in her lab. Her research involves mapping oyster reefs all over Hong Kong using drones and working to restore these habitats. She tells me that the most important ecosystem function of oysters is their filtration abilities; they are crucial natural water purifiers, especially in turbid brackish ecosystems. In one of her YouTube videos she shows a timelapse of oysters in a fish tank, making murky water clear within hours. Other scientists in her lab have calculated that a reef about 7 square metres large filters water the volume of an Olympic swimming pool in a single day. This made a strong case for their restoration in the coastal wetlands of Hong Kong.

Jay also says, 'Possibly the most important aspect of oyster conservation that we try to draw attention to is their historical importance. Oyster farming used to be a very important part of the local culture, but this has been lost over time due to overfishing practices and land use changes. We try to drive the message that we are searching for these "lost" reefs and are trying to restore an important part of culture for the local fishing communities.' This is very much true of India too, in places like Edaiyanthittu.

Over 85 per cent of the world's oyster habitats have vanished over the last hundred years due to overexploitation, bottom-trawling and other destructive fishing practices, pollution and conversion of coastal wetlands into real estate. Conservation and restoration of oyster reefs for their ecological value and crucial ecosystem services is virtually absent in India, where oyster reefs still exist. These habitats need to be mapped and the life they host, unique to the coastal wetlands of India, needs to be documented. They are shoreline protectors and water-health keepers. They warrant to be treated with the same importance as mangroves, seagrass, sand dunes and coral reefs—under Indian Coastal Regulation Zone law.

On the way back I discover my right big toe bleeding profusely, brackish punishment for wading barefoot. And then when I unravel one of the fishing nets to see bycatch, a shred of jellyfish tentacle gets tossed up, sticking onto my face under my left eye, burnt in. Like a pirate, I now sport a 2-inch scar whose permanency I am worried about.

Gummalapuram Krishnagiri
5 September 2021

The paddy fields are booming. The frog calls drum on your skull. Monsoon clouds creep above like large dementors from the Harry Potter world in the night sky over Gummalapuram. Aswathi, Nanditha, Rohith, Anooja and I have come to document biodiversity at Bhoomi College's farm and later do nature sessions for their students.

The uninitiated ear at first hears only caterwauling. But wander the fields with a flashlight and you move through an enchanting spatial orchestration. Distinct aural frog realms are distributed across the waters.

Ke … ke … ke … ke … ke … kekekeke—those accelerating staccatos are cricket frogs. They like to sit and call from the open, damp levees. The edges of the fields flanking the forest, where the ground is dense with foliage, are preferred by the ornamented pygmy frog. Short, drill-bit bursts—they cry pushing forward their whole bellies and throats. They like to stay somewhat hidden under the grass and leaf litter. Right at the corner of the four-square field I track down the call of a red narrow-mouth sitting on the leaf of a prostrate shrub, then find a pair nearby in amplexus. Red narrow-mouths prefer these hills and like the forests more than the fields. Their calls—warm burps—are drowned out by other species.

The katakatakatakata of skittering frogs floating in the water provide the thickest texture to the ambience, most of the time. They start calling the earliest, just before dusk. Bend down and see them along the water's meniscus—they look like crocodiles. Deeper into the night the voices of skitterers quieten a bit when they move out of the water and hunt on the paths, when they change their voice and utter a typical anuran ree-bup. Cricket frogs then take over. Calls of white-bellied pug-snout frogs and burrowing frogs were loud and prominent the previous night, after fresh rains. Today they are entirely absent.

An elephant alarm goes off in the distance. Dogs bark in the village. On drier land and laterite rocks the krrrrrrrrrrrrrrrrrr of common Indian toads and the buffalo groans of painted frogs. A vinegaroon prowls on the footpath alongside us.

In human child development, in the period before the age of seven, the young human being is observed to be most impressionable and is a receptacle to the world around.[29] Much of one's identity, selfhood, sense of home and belonging is said to be formed during this time, shaped by the environment, parents,

[29] Maria Montessori, *Absorbent Mind* (Holt Paperbacks, 1995).

peer group and school. I read later about the unique 'sensitive periods' of other organisms as well, frogs among them. When they are tadpoles, frogs learn and imprint on what 'home' is through 'native pond odour'. This was discovered by Sergei Ogurtsov and Vladimir Bastakov.[30] Frogs register the distinct smell of the wetland where they spend their larval childhoods, cell-deep. They will continue to live, breed and hibernate in it during summer. Any other water which carries a different chemical signature is smelt as non-home. Tadpoles born in polluted waters have learnt it to be their home, and don't move away even as adult frogs, and die fast. I think about how for several hundred frogs from at least twelve different species, these few acres of seasonal paddy are a 'native place' for possibly many generations.

All other species have 'sensitive periods' when certain experiences at certain times during the growth of the organism have a potentially lifelong impact. For instance, young birds and mammals imprint on their mothers or caregivers at the time of hatching. Not just that, infant mammals form rapid and deep memories of the smell of their maternal environment and seek it out. Adolescent animals experiencing hardships in their community develop social anxiety or hostile relationships which can persist through adulthood. This rings true of the elephants here in Krishnagiri. The stimuli organisms receive during their sensitive windows can at times be maladaptive in their lives. Especially given that in a world transformed by human action, other forms of life are constantly faced with novel forces and stimuli they have not yet evolved to understand.

An elephant alarm goes off again in a different direction.

Walking through a frog chorus can be a deep-water wash of the brain, tuning it out of its own incessant chatter and tuning it into

[30] Sergei Ogurtsov and Vladimir Bastakov, 'Imprinting on Native Pond Odour in the Pool Frog, *Rana Lessonae Cam*', *Chemical Signals in Vertebrates* 9, 1992.

the ambience. That night my dreams merge into the croakscape around me. In early sleep I notice my chest heaving and falling to the pulse of the ornamented pygmy frog outside my room. Then the mind segues into a deep sleep dreamscape where there is only the night sky and anuran song, and the listener/thinker dissipates. The observer and the observed have merged.

Gummalapuram Krishnagiri
6 September 2021

Krishnagiri is the northwesternmost tip of Tamil Nadu, interlocking into the Bannerghatta Hills in Karnataka. Elephants claim this landscape as much as people do. Every household, every field here is full of stories of human–elephant conflict and interaction. The news from this district over the last month is filled with it: 'Makhna elephant dies of electrocution ...', 'Third farmer killed in elephant attacks in the last three days ...', 'Elephant tusks seized, four arrested ...' and so on.

In the brilliant *Wild and Wilful*, Neha Sinha narrates how elephant herds move through forests along memorized pathways passed on through generations. Create obstructions across these and the animals will protest. They will try to bring it down and claim their right of passage. Neha writes of a group of elephants at Kaziranga that protested against a golf course wall built across their memory lane and broke it down. The youngest calf suffered a haemorrhage and died. In Krishnagiri the trauma the two species have inflicted on each other runs for generations unbroken.

The ground here in Gummalapuram sometimes trembles with granite mining. Elephants are everywhere all the time, even if you don't see them. As clefts in barbed wire plucked apart like cotton strings. As frisbee-sized footprints, crossing the fields and fruit orchards. As snapped boughs, smudged bark. As dung heaps heaving with latrine flies, dump flies, tiger flies, blow flies, bottle flies, dung beetles, maggots, millipedes, worms, fragile dapperling fungi and sprouting seeds. Elephant dung is its own forest of life.

This afternoon we decide to follow a lone elephant's tracks—which had come in at 4 a.m. that morning—into the hills as far as it is safe and clear. Through the guava plantation, across ploughed fields, through gaps in the rows of gliricidia, down step cultivations and harvested corn, along a steep fellside, then by a lake. Here it had cleared bamboo and slid down on its knees about 30 feet by the lake's bund. Then it had headed into the forest. On the hills, we can see elephant trails and gaps in the canopies, their memorized pathways. On the way back we take pictures comparing our feet and human figures alongside elephant footsteps.

Forest edge—
frogs float in
elephant footprint puddles.

Gadilam Estuary
Cuddalore
20 September 2021

Gadilam is unlike any other river on Tamil Nadu's coast. It originates from the streams flowing down the Kalvarayan Hills of the Eastern Ghats, runs west to east like all other rivers for over a 100 kilometres, but when it gets near the sea its mood changes. It breaks out in two arms from the main channel entering the Bay of Bengal. The northern arm joins the Thenpennai River, which flows closely parallel to Gadilam all through its course. Its southern arm curves inland, forms another estuary 3 kilometres south (where the Cuddalore harbour is now located), then turns further inland for another 16 kilometres where it tentacles out into fields, canals and tanks. One of its channels flows into the Perumal Lake. A waterbody that is over 12 kilometres long and 2 kilometres wide at its broadest—a few thousand acres of water transitioning from reeds to lily pads to duckweed and then shallow marsh. It looks like a chunk of river cast away. Maybe centuries ago this lake was part of Gadilam, till the fickle thing decided to reroute and meander again and left a large portion of itself behind.

Smooth-coated otters are everyday sights, the local people say. Vijay and I see six of them porpoising in the steel-grey water, one backstroking to see if the others follow. My eBird list at the waterbody is sixty-six.

The sand at Gadilam's blocked mouth is black—jet-black like a jungle crow, with organic matter and magnetite (and maybe spilt coal from cargo ships)—and hot as hell under the noon sun. Sand between my toes pricks and burns. Vijay and I are here on MNS's coastal documentation project along Tamil Nadu's shoreline—documenting threats to what have been declared as Important Coastal and Marine Biodiversity Areas along the coast.

We spend the afternoon wading in a creek covered with oyster reefs—mostly Indian backwater oysters, but amidst them windowpane oysters, rock oysters, pearl oysters, green mussels and other bivalves. Oysters clap once in a while to exude excess water. Clap ... clap ... clap ... clap ... clap. These mudflats ring like bursting bubble wrap. Popcorn creek. The receding tide uncovers red lugworm mounds made of fine grey silt. A terek sandpiper buries itself till its neck into a mound, taps its feet and reels out a worm much longer than itself and flies with it into the mangroves. A dozen things squirm, move and dart within the mouths of oysters. Two women nearby collect the large ones and scoop out their flesh into straw baskets on their backs.

Across the reef are about 600 plovers—mostly sand plovers, some little ringed and Kentish. A few whimbrels, stilts, sandpipers and redshanks are around the plover flock. They crouch together and move like minions, from *Despicable Me* scattering out further away. Plovers on oysters are like stones on boulders. Sometimes it is difficult to see even hundreds of them perched in plain sight. Two brahminy kites come quartering close above with attentive eyes, the brown wings made red by low-angle sun. The waders find them more terrifying than me, taking off and blanketing in my direction, making a sound like sand thrown in water, then settling all around, then scurrying to keep a safe distance.

One brahminy flies overhead, circles and yells. About 200 plovers take off and begin to magic-carpet in the air. The flock origami-folding, fluttering, all orienting in one way to show their sandy backs and suddenly vanishing into the oyster-bed backdrop. They then abruptly reappear 40 degrees away as they swivel, teleporting from place to place in the estuary by rotating shades. During a close approach by the raptor, all of them bank together. The sun catches their synchronized bellies and a huge sheet of white flares. A mirror shines into the predator's eyes and mine, stunning us for several seconds. The kite gains height, soars, uninterested. The plovers settle and forage among the oysters.

Pamban Island
Ramanathapuram
25 September 2021

Around forty boats come in from the sea, their nets brimming with sardine (*matthi* in Tamil). Fishermen pour the fish on the ground, stretch their nets tight and yank them up and down. Fountains of fish splatter around. The stuck ones thus get dislodged. Brahminy kites catch them in the air. Vijay and I are hit several times by flying sardines while trying to speak to the fisherfolk at Pamban Beach in Ramanathapuram. Bold kite feet miss my eyes and ears a few times.

Brahminy kites need fish and clean water. I stopped seeing them commonly in Chennai years ago. You may see them in inland lakes, but they are at their best near coasts and mangroves. Their numbers increase as you travel south along Tamil Nadu's coastline. Here at Pamban, these birds are bold as bulls. Even as baskets of matthi are being loaded in ice vans, brahminies plummet down and pick them off fresh. They gather on lamp hoods and trawler masts, swooping down one or two at a time. They make mock strikes, create misdirections while others dive in for the loot. They cry threateningly from their perches—you can see their quivering tongues. They make the men rush to get all the fish inside the ice vehicles quickly.

This cape juts another 30 kilometres east–south–east into the ocean up till Dhanushkodi's *arichal munai* (tip of erosion). To the south is the Indian Ocean at its wildest, raging and thrashing on the sand and on the rocks of the seawall. Immediately north is the Palk Bay, stretching calm, blue and ripple-less. A sight that can only be seen at this point on the entire Indian coast. Right from the nose of Nagapattinam through Thanjavur to Pudukkottai up till this point at Dhanushkodi, the ocean is lake-like, wave-less. It is even called *eri kadal* or lake-sea within the Palk Bay. This end of it has large flocks of grey plovers, turnstones, sanderlings, stints and curlew sandpipers turning into breeding colours. Here the Vaigai River meets the ocean. Its estuary is a water-bird haven. Women wade back carrying buckets of harvested seaweed on their heads.

These two vastly different marine spaces are parted by the dark, sunken forms of the Adam's Bridge running southeast up to Sri Lanka, which is just 18 kilometres away from the promontory I stand on. I go too much to the edge and get a welcome message to the neighbouring country from my phone service provider. These submerged limestone shoals are geo-mythological mysteries. There are many debates about their origin. They are variously identified as fragments of continent strewn across when India and Sri Lanka tectonically separated; sediment accretion from the opposing longshore circulations in the Palk Bay and the

Indian Ocean; fossilized coral reefs; the world's largest tombolo from the thinning of the earth's crust; etc. But the most popular origin theory is non-geological—of the Hindu god Rama and his *vanara* army building a bridge by throwing stones into the sea in an attempt to rescue Rama's abducted wife, Sita. Other rulers and politicians have perhaps found it symbolic to build their own bridges here. The British built one for trade, which now lies in ruins. Indira Gandhi built one adjacent to it in the 1970s which is now used to cross from Mandapam to Pamban Island. Now Narendra Modi has begun work on India's largest hanging sea-rail bridge in the richly coralline Palk Bay waters.

Kovalam / Muttukadu Creek
28 September 2021[31]

The air over the groyne begins to click and crackle. I am at the Kovalam Creek surveying the coast when a swarm of several thousand wandering glider dragonflies (*desanthiri thattan* in Tamil, describing a dragonfly that wanders across nations) envelops us. We stop our work and stand watching. Inside the dragonfly cloud, their wings make electric sounds and radio static as they dart around, landing and taking off.

This cloud of odonates has moved in from the sea. We can see the swarm extend far over the waves. The insects gradually move inland to the tail of the groyne and further. After some time, another wave of insects comes. Then another. Curiously, if you watch the individual insects, they seem to fly at random—like Brownian particles with temper and attitude—tackling the sea winds, turning in hitches and knots, sometimes sparring quickly or chasing each other. These dragonflies allow the winds to carry

[31] This passage first appeared in the 20 December 2022 issue of *RoundGlass Sustain* as 'Wandering Glider: World Traveller and Weather Vane'.

them, forming a river of odonates. If you turn a microscope on the river, you will see its molecules moving at random. But you have to zoom out to be able to discern the river's flow. But I am too small to see the swarm's flow.

Palayam Anna once told me about dragonflies from his younger seafaring days. When he took his boat out to sea in the Tamil months of Purattasi and Aippasi (mid-September to mid-November), the ocean air would be full of thattan. If he and his fellow fishermen ran out of bait, they would use thattan. During this season, large fish like perch, mullet and seer roam close to the surface and rocket out of the water to catch these insects.

It felt like I was beginning to understand the movements of dragonflies from their behaviour near the coast. I was wrong. In the natural world, I am learning that the answer to some questions is unknown and to most questions unknowable. Later, I spoke with the renowned dragonfly researcher Vivek Chandran, and he told me that the dragonfly swarms we see near the ground are most likely feeding swarms. But this is not the height at which they migrate vast distances. Wandering gliders, also called globe skimmers, are the insect world's greatest travellers. These dragonflies make a multigenerational migration of 18,000 kilometres in a year, with a single individual flying over 6,000 kilometres. Vivek said that the gliders ascend high up into the sky where prevailing winds are much stronger and passively glide on them.

If low-flying groups are feeding swarms, what are they feeding on? During a pelagic bird survey in September 2018, I was in a trawler boat, tens of kilometres into the ocean, off Chennai's Kasimedu harbour, when we ran into a massive block of dragonflies. Our vessel was moving quite fast, and the insects striking our faces and limbs felt like needle pricks, so we hid behind the engine room. But what could the dragonflies be feeding on, hanging around so low over the sea? Vivek told me that much of what these insects feed on is invisible to us. They could be feeding on airborne plankton or aeroplankton carried by sea spray. They are especially abundant in the atmosphere when the waves are choppy.

Marine biologist Charles Anderson discovered the best-known circuit of wandering glider migration from his observations in the Maldives.[32] He mapped their triangular route riding on the monsoon winds across the Central Asian flyway—from the Indian subcontinent to East Africa, Central Asia and back. Several bird species—including the Amur falcon, hobby, pied cuckoo, Eurasian roller and blue-cheeked bee-eater—take the same route as the odonates and at the same height of about 2,000–2,500 metres. Anderson suggested that the birds sustain themselves on their flight with insects. But this is just one route that has been studied. Wandering gliders are genetically similar all over the world and seen in every continent except Antarctica. Their migration routes must be a globally interlinked network of circuits, closely in sync with different seasonal winds across the world. In China alone, for example, three migratory circuits are hypothesized.

For a bipedal creature like me, home is on land. It's a circle on the map. For a wandering glider, home is ever the wind, weather and sky. Can that even slightly enter my comprehension? I can only think of the much-quoted verse by the Tamil Sangam poet Kaniyan Pungundranar: '*Yaadhum oore yavarum kelir*' (Everywhere is home, everyone is kin). Was he possibly looking at swarms of migrating dragonflies?

I live in Velachery in Chennai, where odonate diversity abounds, as it was and is a marshland disappearing fast to urbanization. In October, wandering gliders hang by their forelegs everywhere from the dry prosopis thorns to the typha

[32] R.C. Anderson, 'Do Dragonflies Migrate across the Western Indian Ocean?', *Journal of Tropical Ecology*, 2009.

reed heads; they never perch upright. They are known to be rain-pool breeders to avoid competition with other dragonflies and predation pressure from fish in perennial waterbodies. Their larval stage is the shortest among dragonflies, about four to six weeks, tuned to the hydrology of temporary pools. Then Tamil Nadu must be a crucial stopover site for them as it has thousands of eris (traditional lakes) and associated streams, fields and canals which come alive only during the rains. The state has among the largest number of ephemeral waterbodies and wetlands. I've heard a Tamil saying from farmers in rural Chengalpet and Kanchipuram: *'Thattan thaazha parandha thappama mazhai varum'* (If dragonfly swarms fly low, then rain is sure to come). Every year between October and December, weather depressions or cyclones form in the Bay of Bengal, and red alerts are given to Chennai for strong winds and intense rain. Unfailingly, one or two days before a storm, over pockets of marshes, flooded playgrounds, puddles and vegetation, thousands of dragonflies fill the low air. Rain follows soon. I've come to think that the glider swarms are ever-riding the extremities of weather fronts as they hitch from one place to another.

Winds are changing with the changing climate. One phenomenon increasingly recorded in the Bay of Bengal is 'rapid intensification'—weather depressions changing into cyclones in an abnormally short time. Climatologist Roxy Matthew Koll has shown that the land–sea thermal gradient is falling on India's west coast due to warming oceans.[33] This has resulted in more intense weather events but a gradual decrease in total rainfall over the years. Other studies[34] show that the Indian monsoon is becoming

[33] Mathew Koll Roxy et al., 'Drying of Indian Subcontinent by Rapid Indian Ocean Warming and a Weakening Land–Sea Thermal Gradient', *Nature Communications*, 2015.

[34] Neena Joseph Mani et al., 'Can Global Warming Make Indian Monsoon Weather Less Predictable?', *Geophysical Research Letters*, 2009.

more and more unpredictable and that seasonal winds across the world are speeding up. Do the dragonfly swarms carry these messages with them in the ways they appear and disappear? Do they have to fly higher and faster? Do they get time to complete their lifecycles as monsoons turn abnormal? Are their routes changing, numbers falling, wings growing weary? Possibly these insects hold and tell these stories through their movements, and there is much to learn by observing them.

Section 3

RAIN MEDITATION

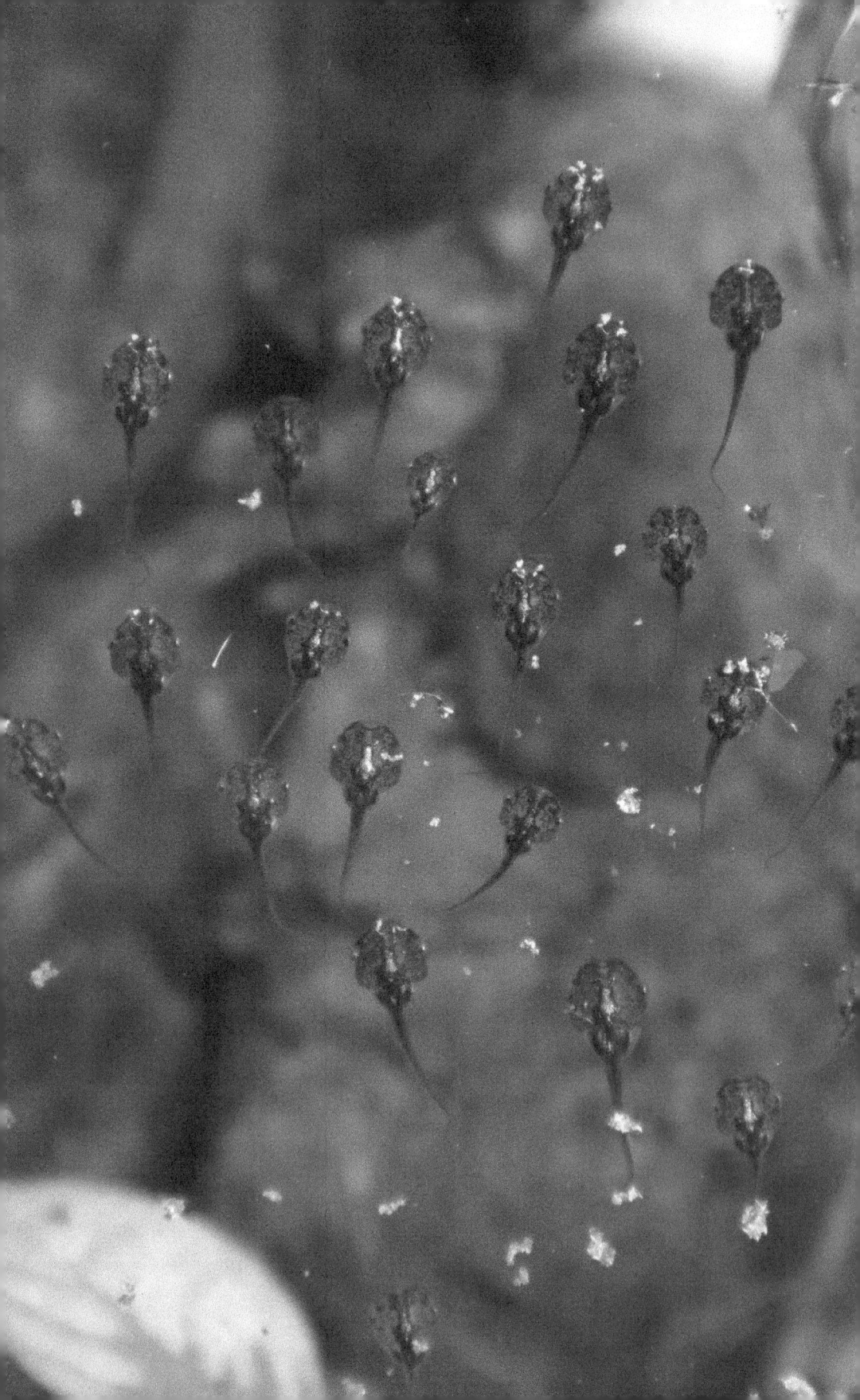

Kotturpuram Urban Forest

14 October 2021

The ocean is in transition now. The longshore currents are reversing direction. Fisherfolk in Chennai say, '*Kadal summa kadakku*' (The sea is lying idle). Some waves curl south, some curl north. Sometimes they don't scurry towards any side at all; they carry forward and break with great momentum, frothing thickly like a mountain stream. Currents sometimes null one another, and the sea is completely silent for several seconds, which perplexes the mind used to its constant sound. Or they collide and leap, like a boulder has dropped. The nearshore waters are brown with churned-up mud.

If it is cloudy, the bee sisters stay in their burrows. Come sunshine and they get busy again. In the fallow land across the balcony, acres of daincha plants are flowering. Common grass yellow butterflies throng in their undergrowth. The bee sisters come back with pale yellow pollen, the daincha's, most likely.

At Kotturpuram Urban Forest, Trincomalee trees (*Berrya javanica*) bear star-shaped pink flowers. They sing and sag with sweat bees. I see giant honeybees come to the flowers, hover around, try perching, then fly away, vexed with the commotion. Sweaties cluster on the flowers two, three, four at a time. Walking over each other, kicking and shoving. They keep boisterous, snuggling company. Some of them bullet straight into a nectaring party to throw a couple off and get space. Other pollinators leave these branches largely alone.

I watch the bees on the Trincomalee for a long while, feeling in my gut the world of difference between us. I hold in mind Catherine Keller's words, trying to understand what it may mean for the bee and me. 'For difference itself is relation: we exist only in the relationality of our differences.'[1] Bees feed me, feed the

[1] Catherine Keller, 'Spiritual Foundations of One World', *Great Transition Initiative*, August 2021.

world. My health is helplessly tied to their tiny veined wings, body hair and the pollen bags on their legs.

Watching the bee melee, I think about the truth of ecology, that too much of one entity devours itself and its own foundation. Differences and diversity of species, perceptions and worldviews are a requisite for life—for existence itself perhaps—which is made of not one but so many fundamental particles, however much we try to find a single god/entity/particle/theory which explains everything, governs everything. This search perhaps tells more of our own mind's constitution, rather than the nature of reality. We keep finding that the 'thing to be known grows with the knowing', in Scottish poet Nan Shepherd's words. The binary mind craves simplicity and single solutions, with its incapacity to hold the ever-deepening lattices of relationality between numerous entities. A different depth of consciousness is required to meet and comprehend the sheer queerness of reality, of life.

Near the park's gazebo, my favourite nectaring plant, *Premna serratifolia*, is in full bloom. I drag a painting stool, stand on it and peer into the branch bouquets in which its tiny flowers are held. They are busy with buzzing and humming in so many faint pitches, with insects landing, taking flight, whizzing around from flower to flower. In no other bush have I seen a greater number of hymenopteran pollinators (bees, wasps and ants). The pollinators I sighted on the premna are:

Greater banded hornet (*Vespa tropica*)
Lesser banded hornet (*Vespa affinis*)
Cuckoo bee (*Thyreus sp.*)
Sweat bee (*Nomia sp.*)
Black bee fly (*Bombilidae*)
Incline fly (*Cylindromyia sp.*)
Thread-waisted wasp (*Ammophila procera*)
Blue flower wasp (*Scolia sp.*)
Bronze potter wasp (*Rhynchium sp.*)

Golden cricket wasp (*Liris sp.*)
Orange-spotted flower wasp (*Scolia binonata*)
Sand wasp (*Bembix sp.*)
Diabolic digger wasp (*Sphex diabolicus*)
Red dwarf honeybee (*Apis florea*)
Grasshopper wasp (*Prionyx sp.*)
Potter wasp (*Delta pyriforme*)
Carpenter ant (*Camponotus compressus*)
Tailed jay (*Graphium agamemnon*)
Tawny coster (*Acraea terpsicore*)
Common crow (*Euploea core*)

Last week I heard my first brown shrike of the season—the harsh, grating call which it voices with such tail-wagging enthusiasm from the barbed-wire compounds near my house. The last sound I have been hearing these days before falling asleep is the soft drilling of ornamented pygmy frogs from the rain puddles. And I am thankful that I can still hear frogs in the city where I live.

Urur Kuppam Beach
1 November 2021

A week ago, the circulation in the Bay of Bengal reversed directions briefly. From Urur Kuppam Beach the ocean now looks like a great river flowing south. On 26 October, Palayam Anna asked me, 'Did you see? The vellam [longshore current] has changed today.' I said no. He said, 'Go look then.' That day was also the onset of the northeast monsoon.

I took my teacher colleagues from Abacus Montessori School, and some of their students, on a shore walk to the Adyar Estuary on Sunday. The winds were blowing distinctly stronger from the north. Mole crab catchers benefitted from the stormy sea; they waded and foraged in the frothing waves and scooped up these creatures from underground, feeling their stirrings with the soles of their feet. Our legs quickly sank into the intertidal, as the longshore current was swift, yanking our ankles.

Here's how to be an anemometer—turn around 360 degrees and tell me when the sand flies into your eyes. Through October, I have been coming back from the beach with my ear cavities, collarbones and hair thick with silt and magnetite.

———

Last night the land roared with frog calls in Velachery. But Indian bullfrogs and Jerdon's bullfrogs have gone missing from here in recent years. I was suddenly gripped with heartache at the absence of their drawn-out groans and lamb-like bleating. Frog call during monsoon weather has become for me a spiritual necessity. It is a source of morning meditation.

———

I see lines of procession ants (*Leptogenys processionalis*) shifting homes everywhere due to rain. They run clutching their white larvae, two or three together, and their long, brown vitamin-pill pupae one at a time. Wherever you see them, the humus is thick, leaf litter plenty and soil full of life. But you see procession ants only where the ground hosts enough termites. These ants are primarily termite raiders, and termites are famous bio-indicators of soil fertility. Tread accidentally on these ant files and they swarm over your leg like mercury or Venom, the symbiote. When they bite, their bodies fold up or lift into the air by the force of their mandibles.

At the Abacus school farm in Vellaputhur, which means 'village of pale termite mounds', I often see the longer and larger slender-jawed procession ants (*Leptogenys chinensis*). They are more difficult to see and they live on the go. They take up rubble space and rock piles as temporary nests, heap up their brood in an interstice and stand guard. Recently, when I was scouring through a brick pile at the farm to find Whitaker's geckos, I found a makeshift home of slender jaws. They took their brood and scurried 2 feet away in the rubble and piled them there, several hundred larvae and pupae vacated in seconds. I picked up the bricks nearby and they vacated again, going deeper in. During the rains, they seek above-ground spaces as their labyrinths underground get inundated.

I found a note in an old diary from Pathashaala, a residential school where I taught earlier, of an April day when procession ants were swarming over the roads of the campus. It was a searing hot morning with weak wisps of clouds in the sky. This time the swarming was definitely not an indication of or response to rain. Close to noon, millions of ants started marching out. Shifting serpentine lines frenetically over all the paths, carrying their white and brown broods. Mostly procession ants, but also a few other species. There were long files everywhere, as though every single nest was being emptied. Ants gathered on classroom walls, stone benches, tree trunks, solar lamps and any other place they found above the ground. Bicoloured arboreal ants congregated on the metal gate at the entrance. It was certainly an ecological omen of something. We couldn't figure out what it was, but soon we got to know that a massive earthquake had hit Sumatra and Indonesia. Around late afternoon all of us felt the passing tremors.

Tadpole Tank
3 November 2021

Diwali is tomorrow. In the rain puddles, frogs, their eggs and their tadpoles are coming to life.

Frogs breathe through their skin. Firecracker powder seeps directly into their veins when it contaminates water, poisoning these crucial wetland health keepers, mosquito-eaters, midnight-singers.

Sharp in my memory are the words of Kanniappan, a farmer friend in Kanchipuram, 'Never drink from a lake which is silent.' When you can hear the frog chorus during monsoons, it indicates that the lake isn't receiving pesticide run-off.

My calves are sore from squatting over the collapsing cattle-feed tank near home. It's now an aquatic algal jungle. Pygmy frog tadpoles dart around; many of them have hind legs now. They gather at the tank's corners, avoiding the centre. I saw a large checkered keelback sliding out when I arrived, the concrete hissing under its belly. Most of the tadpoles had long, mysterious threads hanging from the base of their tails. Poop trails, I thought. Amphibian indigestion for some reason? Later I read up and found them to be cords retained from their eggs. A zoologist friend also told me that tadpoles have very long, coiled guts to increase the digestive surface of their intestine. When I asked her if they reeled out their guts directly into the water when their mouths got tired of eating and drinking, she seen-zoned my message.

Tadpoles part. A nymph of a wandering glider dragonfly climbs out of the water onto a stem of an alternanthera plant and ecloses. I find myself singing a Malayalam nursery rhyme on the 'Thula thumbi' as it does. Tail curved, wings wet and stuck, it looks like a giant lacewing as it exits its old skin. Then it stretches itself into a dragonfly over the next half an hour. Nearby, a granite ghost ecloses too.

A ragpicker comes along and asks me, 'Anna, enna meenu natthai pidikkiringala?' (Brother, are you catching fish or snails?). I show him photographs of the tadpoles and point out a gang which has gathered in a corner. He peers in, picks up the plastic bottles inside the tank and drops them in his sack, wishes me again and takes his leave.

Nearly all day today the sweat bee sisters in the balcony pots have been inside their burrows. They are averse to travel during overcast weather, but perhaps also they smell the smoke, hear the sounds. I see them briefly hovering over the flowering jasmine at 4 p.m., and they go back in before 5. The first frog call I hear is the cricket frog's accelerating ke ... ke ... ke ... ke ... ke ... kekekeke at 4.30 p.m. The rest join much later in the evening. Eight glossy ibises swerve clumsily in the sky as crackers go off on the ground. A flock of cattle egrets nervously scatters, regroups and flies away in a different direction.

Thiruvanmiyur Beach
5 November 2021

The ocean is pale blue like the Andaman Sea today, extraordinarily still. Its waves are a mere bobbing on the sand, crests only about a foot tall. Each cloud of sand and shred of trash seen clearly as the water curls. I walk along Valmiki Nagar Beach to see what the fisherfolk have brought. One man lounging under the shade of his boat tells us, '*Netthu ellan thanni adicchittu padutthutanga sir, kadalukku yaarum pola*' (Everybody got drunk last night, no one went to the sea).

Days such as these are called *thelivu naatkal* by fisherfolk (literally 'clear days', but referring actually to the light blue ocean, calm and clear). They despise them because the fish can see you coming from a long distance and scatter away. The fishers here crib about the fisherfolk in Kovalam during this time. Kovalam is 30 kilometres south of Chennai and it has a rocky ocean bed, unlike Chennai's sandy one. On thelivu naatkal, much of the fish from these waters move to Kovalam where they can hide in the rocks and corals. 'And the Kovalam folks become rich when our nets come back empty,' folks here say. They prefer a rough sea any day to a calm, clear one.

I meet a man casting hook and line from the beach. He shows me how he attaches the meat of wedge clams and mole crabs to the metal and throws the nylon line into the water. In minutes he catches several *keechan* (small-scaled terapons) roaming right within the waves on the intertidal.

Still air keeps the apparitions of Diwali smog from dispersing near my home. My mother and I develop mild fevers. The air quality index (AQI) goes up to 385 (hazardous) from around 100. Noise goes up to 79 decibels. Heavy metals in the air, including lead, nickel, arsenic, barium and aluminium, are measured by the Pollution Control Board to be well over safe levels, rising from 'undetectable' in the previous days. The streets are carpeted with yesterday's blasting and burning. I speak to two corporation street sweepers on the road where I live and ask what the streets look like. One of them says, '*Pattasu kuppai perukki perukki mucchu adaikidhu*' (I am unable to breathe after sweeping firecracker waste all morning). The smell around the rickshaw on which they carry the sweepings is thick and suffocating. I feel it is only right that people bursting crackers have to be responsible for cleaning the mess they create on the roads. Firecracker powder is classified as 'toxic hazardous waste'. The little swifts nesting in the train station towers—usually ribboning in the sky by dawn—do not come out till late afternoon. I suppose the black kites and pelicans which fly south to the marshland every morning have decided to stay at their roost till this dangerous festival ends, as they are entirely missing.

Thiruvottiyur
North Chennai
6 November 2021

This expanse of ocean stretching along Chennai's coast is sometimes called *moonu adukku kadal* (three-stepped sea) by local fisherfolk. For most of the year one can see waves breaking along three distinct planes in the nearshore waters because of the unique bathymetry. But during stormy weather, like today, waves break along five or six bands starting far in the ocean. Three friends—Anooja, Rohith, Sudha—and I are on a groyne at Thiruvottiyur. Sea thrashes the big rocks from both sides and turns into a salty haze. Far out in the northeast, grey pillars of rain stretch down from the crowding clouds.

We have come to Kaladipettai Government School to see if it is possible to create a coastal curriculum for the children. Skill-building around literacy and numeracy can be linked to their home beach—something we are putting into practice in South Chennai. But realities obliterate the proposal. This neighbourhood immediately north of the Chennai port is one of the highest eroding areas on the Indian coast. Between Thiruvottiyur and Pulicat—a 30-kilometre stretch—the sea enters/erases 3–50 metres of land every year, triggered by large coastal infrastructure. As you drive down the ocean slaps away right by the roadside without any buffer. It is quite an unsafe coast to walk on. Geeta, the headmistress of the school, tells us stories of this landscape's past. When she was a child, her grandfather, a fisherman, would take her walking on the large beach which was a kilometre wide. Her small feet would begin to ache before she could actually wet them in the waves.

Beaches are made by rivers depositing sediments at their mouths over millennia, which are ground down by waves and distributed by longshore currents. Thiruvottiyur is in the floodplains of the Kotralayar, Chennai's largest river—about four times as big as the Adyar and Cooum rivers put together. Geographically speaking, North Chennai would have had very wide beaches if there had been no anthropogenic activity.

The headmistress said that in 2006 her school had 450 children. She recollected hearing a bomb-blast-like sound from the shore one day. All the teachers and students ran to see what it was. A massive section of the road had broken and collapsed into the sea. After this incident a number of villages had to be resettled on Kattupalli Island and the school strength reduced to less than a hundred.

The Kaladipettai school is locally known as Mun School—*mun* in Tamil is mud—referring derogatorily to the children there as dull. A fast-eroding coast impacts children's learning and is woven into every aspect of people's lives and stories. Identities, ancestral beaches and childhoods are also being eroded. I ask Geeta, ignorantly at the time, if there is a beach at least somewhere nearby where we can take children on shore walks. She says, 'Most of the children here haven't experienced a beach growing up. They think beaches are there only in the posher neighbourhoods of South Chennai.'

Tadpole Tank
13 November 2021

Between 5 and 5.30 each morning I sit to meditate. My practice has been different this last week. I sit on my mat facing the open window, close my eyes and merge my attention with the frogscape outside. I scan it slowly from near to far, side to side. When thought rises in a different direction, I gently bring it back to the sound of frogs. I move my attention to different calls and rhythms and stay with each for a while. I drop the names and knowledge and associations and just listen. I listen to the pauses between their calls. Then I listen to all of them together and let that rich sound wash over me. Towards the end I touch the ground and give thanks to these life forms, for their presence and the spiritual connection they offer me. They have been waiting for the rains to break their sleep and to emerge from the undersoil.

The flood waters have receded significantly after four days of intense rains. I wade through the street to the tadpole tank a couple of times in between rains to see what the weather is doing to it. I find clutches of ornamented pygmy frog eggs floating on the water or tethered to leaves and plastic debris. Each egg is shaped like a red blood cell, a translucent, biconcave disc. On one visit I find two adult cricket frogs in amplexus in a corner. A checkered keelback has coiled like a jute rope around the alternanthera stems, and tadpoles swim all around it, unafraid. Tadpoles with growing hind legs are faster and more fidgety. They dart across and dive down in a blink if a shadow moves. A common tree frog's spawn hangs from a wireweed over the water. At the Abacus farm school where I teach sometimes, they spawn under the lid of the drinking water drums if somebody leaves them open. And occasionally a child yells that there is fish coming from the tap and swimming in their tumbler.

———

Frogs live where land is open to sky. They move up and down with the water table. When cement, steel and tarmac arrest water above and shut off the soil beneath, I wonder what happens to them. Do they die en masse in their sleep, unawakened by the percolating moisture? My friends living in areas at the heart of Chennai, in totally paved places like T. Nagar, even though their surroundings are flooded now, tell me that they are having silent nights.

My frog-meditation sessions each leave me full of new perceptions and observations. I hear each species calling at different pitches, and I can move my attention through them— as if from cellos up to flutes in an orchestra. I also notice that each species has a unique and different tempo. That intrigues me because this difference allows for pauses during which other voices can be heard, into which other species can fit their phrases. A white-bellied pug-snout frog almost exactly matches the clock, at around sixty calls per minute. A painted frog's longer groans with lengthier pauses are about thirty to forty calls per minute. Its voice resonates within my head and runs volts through my brainstem, leaving me tingling and totally present. Cricket frog phrases are about 3–4 seconds with 1–2 seconds of pause.

If you listen closely to a frogscape, what stands out is how the different calls alternate, of one species tuning into another's phrase gaps and pauses. Looking this up I find the work of Gary Rose from the University of Utah who discovered that frogs have neural circuits which let them 'count'.[2] And that breeding success is greater in males which keep a better tempo. If keeping beat is an innate aspect of these animals' existence then a rainy night really is a chorus. Each frog is always deeply listening to, learning from and participating in the collective sound.

Passing storm ...
the missing pieces
on a dragonfly's wings.

Rain Meditation
Velachery Marshland

I sit on the balcony watching water flake down like snow from churning dark sheets of clouds. It had rained all night and through the morning. The reverb of pug-snout frogs was louder than all others. It had lingered in my dreams and into the early morning. *Aen ... aen ... aen ... aen ...* single notes beating in and out of phase, transformed into a dreamscape of talking and laughing elementary school children, and I am sitting with them, as if among friends. Early in the morning a toad calls *krrrrrrrrrrrrrrrrrrr* from across the fallow land. A magpie robin

---

[2] Gary J. Rose, 'The Numerical Abilities of Anurans and their Neural Correlates: Insights from Neuroethological Studies of Acoustic Communication', *Royal Society Publishing*, January 2018.

sings in the rain, its notes sounding remarkably similar to the first phrase of Vivaldi's 'Theme of Spring'.

With the first light, two Blyth's reed warblers *tcchh tcchh tcchh* away in a prosopis bush below my balcony—small winter visitors from thousands of miles away. Cattle egrets are surprisingly white in the dull weather and a golden oriole crossing the horizon is radiant against the overcast skies. There is a splay of wet termite wings at the doorstep when I go to get the milk. A gap in the clouds moves across like a carriage window over the train station. Then there is an explosion of pigeon wings from the train station's rafters as black kites begin to fly low towards the Pallikaranai marsh to the south. Openbill storks are moving from their roost, not in a flock but in a loose scatter, along with the kites in the drizzle. I write and practise a rain meditation evoked by the morning's soundscape and the peace it brings.

(Rain meditation to be read out slowly with sufficient pauses between sentences and longer pauses between paragraphs. To be practised indoors during rain. Different people's rain memories can be shared at the end.)

Gently close your eyes and listen to the rain. Let your awareness merge with its sound. The way it falls on the roof, the windows, the trees, the ground. What are the different surfaces on which you can hear the rain drumming? Listen keenly, curiously. Scan rain sound beginning at the space in front of you … slowly moving your attention towards the space on your right, making sure you scan and listen to every inch of space in between. Pause in places which make you curious.

Gradually move your listening towards the space behind you. Listen to all the surfaces responding to rain's falling. Do different leaves, different trees sound alike or unlike as rain falls? Do you also hear other creatures awakened by the rain? Frogs, birds, crickets, cicadas? Can you hear the wind through the rain?

Now gently move your listening towards the space on your left. Listen if the rain falls with the same intensity. Is it growing louder, softer or does it stay the same? Does the sound of the rain wax and wane?

Slowly move your attention again towards the space in front of you and the entire space above you. How do human-made things sound in the rain—walls, windows, vehicles, footpaths, other surfaces?

Listen keenly for a minute to the rain sound all around you ... As you listen, imagine the rain in the clouds falling on the soil, on the hills, filling rivers, streams, ponds, lakes and puddles. Imagine it waking up the frogs sleeping deep under the mud, the crickets, moths, fish and numerous other creatures. See the rain flowing up the stems of plants, trunks of trees. Then into other creatures and into you. Returning to the sky, forming clouds, falling back as rain. See yourself in the water cycle, linked to everything through flow and evaporation and falling.

Now as you listen to the rain, enter your memories. Can you scan your mind for a powerful and happy memory which rain evokes whenever you want to remember or relive? Can you bring that memory back now? Where was it? Who was there? What did you feel? Let rain sound help you relive it now.

Gently bring your attention back to the rain around you. What does it smell like? How does the air's touch feel on your skin? Immerse yourself in the sound of the rain.

Velachery Marshland
17 November 2021

Ribcage sky. Exactly between 5.50 a.m. and 6.10 a.m. the brown shrike starts calling. Its voice is not musical in the accepted sense; it's quite scratchy in fact, but I find it pleasant. It flits from fencing post to post and calls with childlike jubilance each morning. Daybreak is reason enough to be excited daily. I might miss the bird when it migrates at winter's end. I heard it first this year on 7 October at 10.15 a.m.—I have this jotted in my diary with a cringy number of exclamation marks.

Intense rains are predicted for north Tamil Nadu tomorrow; Chennai has been given a red alert. The skies were empty for the last four days but yesterday afternoon another wave of wandering gliders began congregating over flooded playgrounds and pockets of marsh. I expect their numbers to increase today.

Ornamented pygmy frogs in the tadpole tank are now turning into froglets. They spend most of their time with their heads out of the water, breathing air. Each frog has a unique patch on its back that makes it easily distinguishable from others (like a tiger's Wang sign). The mark on one looks like a bunny rabbit with long ears. On another it looks like the Pillsbury wheat man. A wolf's face, an octopus, a stretched palm, an inkblot. I look from frog to frog and froglet to froglet, excited like a shrike that I can tell the individual frogs apart.

It hasn't rained for a few days, only the cricket frogs call at night. Nearly all other frogs living in my surroundings fall silent. Which makes me wonder—do different frogs prefer different levels of wetness and humidity to vocalize? Or even to awaken from their underground sleep?

An orange lynx spider ziplines onto the water's surface from a leucas plant, misses something—a tadpole or a water skater—and rappels back up. The leaves of the same plant shake because of a scuffle between two damselflies. A coromandel dartlet has grabbed a pygmy dartlet by its neck. I notice for the first time its barbed yellow damselfly legs as they cage in the smaller damselfly. The coromandel dartlet chews the pygmy dartlet from the head downwards to the thorax. Ten minutes later it drops the pygmy dartlet's tail and wings into the water and licks its barbed feet. A water-skater finds the devoured remains.

Nearby on the discarded steel from the railways a garden lizard basks, its skin much redder than the rust.

Tadpole Tank
20 November 2021

Look into the water too closely and a backswimmer will stretch into a deep-space kayaker rowing through emptiness, paddling between galaxies. Or it will shrink into a tiny, drifting particle inside the liquid of your eyeball, ever evading your gaze. I think I went into a trance watching them today. What are these insects? Surely they are not of this world. Effectively two-limbed, its other four tucked away on its chest so you'd never know. When resting it holds sometimes on to the water's meniscus like the surface tension is a grid of ropes. It watches the sky all day and backstrokes. And backstrokes. A classic space boat with two long oars. Hundreds of millions of years old.

In the cement tank every morning for the past few weeks I have seen at least one granite ghost dragonfly eclose from its nymphal skin; being born a second time. I see tadpoles swimming together and feeding near the walls even in battering rain. This is me observing from outside the water. What is rain within a water body? What might it be felt like? A rhythm of temperature? An expansion of space? A water-skater rests on a clutch of apple snail eggs till a common bluetail damselfly dive bombs, chases it around and makes it scamper into the vegetation.

In the night I dream about walking along a marshy, unfamiliar beach, seeing shells in colours that don't exist. I wake up with the sea's roar still in my ears.

All through this month I've been doing sessions on frogs and wetlands for every kind of age group. With primary schoolchildren (aged four to six) we did a module on 'Frogs of Chennai'. We saw the lifecycle of the ornamented pygmy frog—from egg to different stages of the tadpole to adult—in pictures I had taken in the cement tank. We learnt to identify common frogs in the city and their calls. I explained the role of frogs and toads in their ecosystem and how they are extremely sensitive to environmental changes as they breathe through their skin. We learnt how to draw a common Indian toad, step by step. The children's assignment was to visit a marshland nearby and spot at least five species of frogs and toads. And listen at night to frog calls, notice which species call the most in their neighbourhood, how the sounds change before and after a rain spell.

Their teacher Sudha later called me about the impact of the session. The students had gotten curious about these creatures and spent rainy evenings listening and recognizing calls near their homes. One child had brought a toad home in a cardboard box, given it a name, called it her friend, and perplexed her parents. Sudha also told me about a child who earlier was not particularly interested in learning to read or write. Frogs, however, gave her a new purpose for language learning. She had sat her parents down and had them read out things about frogs, and made them help her read and understand and learn new words. In school too her teachers supported her interest. I witnessed her reading a few months later about a frog's life cycle with rare eloquence for a five-year-old child.[3]

Working with primary schoolchildren, I began to think there is—there must be—a deep missing, a yearning at the core of our ecological selves when frogs and crickets and songbirds go missing around us. Perhaps even if a child comes into the world never having seen or heard frogs or magpie robins, some part of the baseline may never shift and is unrestful. Because our cells formed through aeons hearing frogsong, birdsong, cricket song.

[3] Yuvan Aves, 'Can Wild Places Create Equal Learning Spaces?', *Vikalp Sangam*, June 2022.

Songlines School Farm
Vellaputhur
4 December 2021

Wild banana in Indian jungles is a sight to behold. If you drive past them in the Western Ghats or in the Northeast, you see them cover entire hills, form their own forests. Banana wild is like fungi, like octopus, like angler fish. It resists entering our language and comprehension. A whole hill of bananas could really be just one vast, ancient being. If you've cultivated banana, you'll know. New plants sprout from the corms of mother plants underground. But mother and child are our words, not theirs. They grow from the same root, connected. In Tamil '*vazhaiyadi vazhaiyai*' (like generations of banana plants) is said as a blessing to describe families which have held and lived together for generations.

If you've desuckered banana, you'd have seen that its stems spout water. Water flows out of its severed rhizomes like a spring. Imagine a banana forest—underground it is a network of rhizome streams. An alive hydrology that other vegetation and lifeways tap into. Bananas in the wild are known to accelerate the regeneration of natural forests.

Mist set in yesterday over Chennai. The mornings are cooler by several degrees. At the school farm we raise vegetable nursery beds to be transplanted into the fields in January. Numerous froglets hop on the slushy paths. After sowing seeds, I go with Class 9 students to desucker half an acre of *karpuravazhai* bananas. Red dwarf bees (*Apis florea*) have a hive on the corner-most plant. Sit by and listen to a beehive for a while and your scalp starts to tickle and your ear canals begin to twitch with their murmur and movement.

On the tree right next to the hive, one of the last banana flowers hangs from the topmost stalks. Greater banded and lesser banded hornets are hording to it. They gather on the fertilized female flowers, which are turning into firm green fruit at their base.

I keep going to see the insects between noon and quarter past three, till I leave the farm. I leave the anchoring of the children's activities to my colleagues.

These flower bunches are hornet water holes, with some bottleflies and other minuscule flying insects coming to them. The hornets reach into the rear of the flowers or hang behind them—in rows of yellow and black uniform—chew on the top edges of the tepals and drink the juice oozing out. Some take a quick swig and fly west, where perhaps on some large tree or fallen log there is a nest. Others stand gulping for minutes, some even ten, like they have gone thirsty for days. The larger species mob and chase the smaller ones if they want juicy spots. Is this bunch just a hornet pond? Or is banana flower juice their health drink? Can they now fly faster and for longer? Or do they regurgitate it and mix it with chewed-up wood to make stronger nests? I don't know.

I've had two close brushes with hornets. The first was when a rain tree was cut down with a nest on it. It was on the campus of the school where I taught. The insects got furious and attacked everybody around. School got a half-day hornet holiday. The second was while trekking in the central Himalayas when a nest of Himalayan hornets (*Vespa vivax*) was trampled by mules and the insects were enraged for days near our campsite. However, away from their nest and unharmed, these feared insects are quite docile and just go about their business. I take pictures for hours, inches away from them. One hornet lands on my phone's frame, licks its forefeet thoroughly and flies off again.

Nearing storm—
the bamboo gate whistles
through bee holes.

Section 4

DETRITIVORE MEDITATION

Kotturpuram Urban Forest
13 December 2021

Ten of us sit around a tall compost heap at Kotturpuram Urban Forest,[1] filling packets for seedling trees. Each handful crawls, writhes and bleeds with life. A rotting rain tree twig crumbles in someone's hands. A whole village of bicoloured pennant ants (*Tetramorium bicarinatum*) pour out of their apocalypsed rooms, pantries, crèches and halls. Major workers, minor workers, alates and all—about a lakh and fifty thousand of them—carrying their white eggs, larvae and maroon pupae, hurry to find another safe twig or dung cake. Deeper inside the heap is an albino millipede, shyer than its dark-coloured siblings, crawling over bits of leaf, sensitive to sunlit squares, curling to camera clicks.

Kotturpuram Urban Forest is a wild oasis in Chennai, located on the bank where the Adyar River curves like a camel's hump a few kilometres before entering the sea. This place was a debris heap two decades ago. It was reclaimed by Nizhal—a trust for tree conservation founded by Shobha Menon—and gradually turned into a community forest with the participation of volunteers from various neighbourhoods in the city. I was one among the hundreds of children who often came to this place. I first began visiting, to water the trees and add compost for them, when I was in Class 9. I've been coming regularly since then and have been able to bring with me many groups of children after becoming an educator. I remember when the trees were just saplings and the park an arid space with sand, grass and rubble. Everyone had to wait in line to use the one hand pump that was there to fill buckets and water saplings. Now it is a forest.

Our group has three middle-school students, two school teachers, two nature educators, a zoology professor and two life

[1] A version of this passage first appeared in *Sanctuary Asia*, Vol. 42, No. 6, June 2022.

science undergraduate students. A compost heap is a great space for grounded conversations. Air–lungs–blood are thick with twenty flavours of geosmin. Hands are dirty, eyebrows sweaty and fingernails packed with humus. We speak as we uncover with each grip the creaturely labour which creates soil. A young boy holds the fat, squirming larvae of a rhinoceros beetle in his hands for the first time. We find and watch inside cupped hands wireworms—the honey-brown larvae of click beetles—and the larvae of black soldier flies, which can wriggle on their backs just as fast as they can on their feet. Amongst us teachers and college-goers, we ideate lesson plans as we brainstorm all the natural history and science learnings that a heap of compost offers. We spend time cursing indoor classes.

Seeds of putranjiva and arjuna trees are germinating deep inside the heap. A blue-black flower wasp (*Scolia sp.*) lands on it and scouts around while shivering its iridescent antennae. It looks into mats of leaf litter and chunks of dung. A flower wasp, through its presence, speaks for the local earth's health. Mother wasps find beetle larvae, pierce their skin with their ovipositor (egg-laying needle) and deposit their eggs inside the soft bodies. Wasp larvae devour the living tissue, pupate and emerge as adult insects, eventually killing the beetle grub. Good populations of grubs are found only where the soil is rich in organic matter, where enough leaf litter falls to the ground and is left to naturally break down, where all kinds of animals live and eat and poop. Then the flower wasps come around. They indicate that the soil is full of compost-heap creatures, these world-feeders we call detritivores.

The work of detritivores returns good health to habitats, reverses extractive and depleting ecological processes, and makes endings beginnings. They show that beginnings need endings. They let life thrive wherever they walk, crawl, burrow and slither. They are millipedes, earthworms, dung beetles, domino roaches, woodlice, pill bugs and dung flies—their whole life cycles on and under land. Water snails, shrimp, the larvae of crane flies, caddisflies and mayflies in freshwater. Decorator worms, peanut worms, starfish and sea cucumbers on the ocean bed. Lugworms, fiddler crabs, red ghost crabs and wedge clams on the beach.

I have found essential spiritual and metaphysical meaning in the work of these creatures, in my own path towards healing. Detritivores lend their castings as affirmations. They have taught me to discover within myself the ability to turn the trauma and suffering of my difficult childhood into inner growth, strength and wisdom. To turn the shit of life into something like fertile soil. Describing them doesn't seem too different from describing my own story of learning 'to be a detritivore'.

My stepfather died a few days ago on the night of 9 December. He lived alone for the last two years and passed away in his sleep. Only a few relatives and friends visited him. He and my sister had a tremendous bond when she was alive. Me, he despised and subjected to daily physical abuse from when I was eleven till I turned sixteen. During those years I was barely sent to school and I sustained many serious injuries. He came into the picture when my biological father—who had several extramarital affairs and took little care of the family—left us. This left my mother broke and broken in every sense. When I was sixteen I ran away from home to escape the violence and found mentorship under G. Gautama, who was at the time the principal of the school I attended then. He is one of the greatest educational thinkers I know. I found a home at Pathashaala, the new school he had started then under the Krishnamurti Foundation India. I came back home five years later when my stepfather had beaten up my mother, and I demanded that the violence stop once and for all. He left us, not being able to live in the same house with a strong version of me.

I did his last rites despite opposition from his friends who knew of our difficult relationship, because nobody else would do them. I got to know a little more about him when I met his elder brother after the funeral. My stepfather himself went through terrible abuse all through his childhood and adolescence due to a violent and alcoholic father, who also tortured his mother and his siblings. He ran away from his hometown but carried deep psychological scars. And his father too had a parent who was physically abusive to his wife and children. The brother narrated a cross-generational history of abuse, PTSD and miserable families. Though the children hated how they were treated by their fathers, they ended up emulating them and passing on their suffering to their children, without getting an opportunity to heal and recover.

I, however, had a beautiful childhood ensured by my mother and the wildness of my school campus. And I did everything possible to heal myself and end this cycle—Vipassana, daily journalling, psychotherapy and, more than anything, being in nature. My mother and my school encouraged me to build deep emotional and physical connections with the natural world and with all other species. Somewhere within me lay a gift that saw me become part of the vast mystery and community of life. I sought meaning and solace from banyan trees, pied wagtails, moon moths and millipedes. This I know to be the core source of my well-being and resilience—being a citizen not just of human society but of a boundless multispecies community that helped me recognize I was larger than my suffering and let me continue dreaming along many different paths. It is difficult to describe how or why, and I do not simplistically offer this as a way for others. Neither do I want to give the impression that the journey was easy. Trauma endures. Parts of it never leave. But one can restructure its purpose into something fertile—like earthworms and beetle larvae inside the compost heap of our minds. To be able to do so is possibly the greatest of all our life's challenges. I remember how on some of the most painful days, the sighting of a paradise flycatcher, vinegaroon or a rat snake would immediately lift me, repurpose my life and help me 'hang in there'. Sadness or an aching limb would hurt a little less.

It is possible that my drive to work with and learn from children comes from my own suffering as a child. One of my core learnings as a teacher, having worked with hundreds of children across ages and backgrounds, is that every child comes into the world with a distinct, incomparable energy, agency, even direction. No child, even as an infant, is a blank slate. Kahlil Gibran's words ring true: 'You may house their bodies but not their souls, for their souls dwell in the house of tomorrow, which you cannot visit, not even in your dreams.' And that is why the greatest teachers of our times, from Maria Montessori to Rudolf Steiner to J. Krishnamurti, were students of children. They crafted their philosophy based on deep observation and care.

In my practice as an educator I see inter-species observation and connection to living places playing a foundational role in healing young people through the traumas they are born into.

––––––––

We finish work at the compost pile and leave the little bags in the nursery for saplings to be planted in them later. Then we take a walk through the forest to see what creatures the rains have awakened.

Urur Kuppam Beach
19 December 2021

Through many days of December my 5 a.m. meditation practice has been 'ocean breathing'. Not as it is usually practised but in a way I've evolved for myself. I sit relaxed on my mat and close my eyes. I bring my attention to the area of my diaphragm. I breathe in as I distend my stomach and breathe out pulling it back in. As the air moves in and out I make a sound with my glottis like ocean waves crashing and falling on the shore. I visualize the waves crashing and receding within me as I breathe and as I listen to the glottal sound. Breath, all breath, is mostly ocean. I gently bring back my attention to breath and ocean if it drifts away, immersing myself in the ebb and tide, the silence and sound of both. I do about 120–150 breaths and it takes me into a space of deep clarity and stillness.

Today I conduct a shore walk for twenty-seven people. We walk from the Karl Schmidt Memorial on Urur Kuppam Beach to the Adyar Estuary, whose sandbars have shifted south. Four humpback dolphins bless our walk. They prance around the flag buoys and we see them leap thrice, from nose to tail, straight out of the water. They are seizing escapees and leftovers as fisherfolk reel in the trammel nets they had anchored the previous day for catching cuttlefish. The beach berm has become a cliff edge along some stretches, falling steeply by 6–8 feet along the high-tide line. I remember this phenomenon when it happened two years ago; one moment I was explaining something dramatically to some students and the next moment people were searching for me as the sand escarpment gave in and I went down with it. A strong northeast wind of around 13–15 knots has been blowing consistently at daytime for several days now, driving the south-swiping waves to carve the shoreline deep.

The last few days have been *marappu*. Another word that fisherfolk use for clear blue water and a calm sea, unconducive to catching fish due to good ocean visibility. Anand, as he unfurls his net, tells us that on these days he'd rather go for crabs in the clayey parts and by luck possibly catch fish hiding there.

A black kite is behaving like a pelagic shearwater, gliding over the growing swells and tall, plunging waves, being lifted aloft and then gliding down again to find the next swell. Touching or almost touching the wave's forehead with the curve of its talons. Breezy ocean is no place for a mobbing crow, which tries flying in and clutching at the kite's back. It gets tossed over by the wind like a tumbleweed till it comes back and sits on the sand. I see two northern shovellers over the ocean flying in and banking towards the estuary. My first shovellers for this season, and perhaps the first pair to arrive in Chennai this winter. Brown-headed gulls fly one by one along the swash zone upwind. Little egrets hunt clams. But the kite is obsessed with its game with the waves, discovering the different things its wings can do with draughts of sea wind.

The longshore currents this week have carved the coast in the shape of a sickle near the Urur Kuppam dockyard. It protrudes far into the sea by the turtle hatchery. The intractable, unruly, shapeshiftingness of the shoreline presses on me today, leaving my thoughts turbid and inner weather gusty.

Urur Kuppam Beach
22 December 2021

It is colder in Chennai than my skin can believe. And I am thinking about all the irreconcilable things of the world. When you encounter them they can linger in you for days. Carefully constructed meanings begin to show cracks, and you find your worldviews worm-eaten.

Writing about the sea ...
my finger scab stained
with cuttlefish ink.

———

This morning I sit on the beach with Arumugam Anna, a fisher elder from Urur Kuppam, learning a bit about the geography of the open ocean stretching out from the Adyar River. He is a seasoned fisherman of six decades, yet he speaks with a near reverence for the unknown and uncertain, as if it were a creature of the ocean. When I get too pinpointed with my questions, he says, 'We can never say that for sure' or 'Each day is different' or 'It depends on the sea'. This is me trying to understand the taste of a fruit without having held it or bitten it.

Four nautical miles from the river mouth is a portion of the Thiruvanmiyur kallu. Here fish were found roaming in *kalakku* (mixed species schools) today. He shows me a large *kal navarai* (yellow spot goatfish) and a dozen *kal sankara* (blue-lined snapper) he's caught. Then he begins to tell me the names and locations of the seru (clayey or slushy ocean bed areas) which are good for fishing. *Nadu* seru, *thamal* seru, *kel* seru—which is just beyond the underwater hills—*kalama* seru, *pann* seru and so on. His descriptions and 'mind map' of the ocean across its dimensions begin to flip my brain. Sea and coast, wind and water current are part of Arumugam's lived selfhood like it is that of an olive ridley, humpback dolphin or the whiprays his sons have caught that day.

I pick up a strange slipper near the high-tide line with several white barbs pierced into its heel. Arumugam yells at me to fling it away at once. I was holding a discarded slipper which they had used to affix the venomous ends of whipray stingers they'd broken from beneath their tails.

The few boats which come in are carrying trammel net heaps and are covered in black cuttle ink. But most of the men from Urur Kuppam have gone to Injambakkam about 12 kilometres south as a supercolony of prawns has been found there, and the village sent word to all the others nearby.

One fisherman whacks aside with a stick a marbled octopus which is trying to enter a crevice in his boat while he is cleaning it. Two or three more are crawling around the fibreglass surface trying to find water. This scene leaves me disturbed, seeping into my thoughts. I have read too much about the sentience of cephalopods—in *The Soul of an Octopus* by Sy Montgomery and *Other Minds* by Peter Godfrey-Smith—and then seen *My Octopus Teacher*. For the fisherman they are merely intelligent vermin, bycatch. And he is my friend. How does one speak about conservation to those who live day to day, meal to meal? I see my capacity to think and communicate, to reconcile binaries as nascent. I feel morally jumbled up.

I see a mirror shine in my side vision. A flock of sand plovers banks as it flies over the beach, catches the sunshine with surf-white bellies and swivels them at us. It settles on the ipomoea creepers and a sheet of garbage—plastic cups, glass bottles, coconut shells and thermocol chunks—the birds merging, invisible in both. The sight of a plover flock breezing lightens the day's heaviness.

I go to the neighbouring village of Odai Kuppam, where fishers stretch out the four ends of their yellow sardine nets and toss the fish onto a tarpaulin sheet. The friendliness of artisanal fishers has been the case almost all across the Indian coast. At Odai Kuppam I ask why the men didn't go for cuttlefish that day; it is a thelivu naatkal, a day with clear blue water, and fish are difficult to catch. They feel that trammel nets, which need to be anchored one day ahead, waste too much. Lots of fish other than cuttles get caught, die, rot and are eaten by other sea creatures. 'We might as well go deeper looking for sardine, though the catch won't be great,' Surya, a young fisher about my age, tells me. They find some shoals at the intersection of the *kalakodu thanni* (muddy current) and the *thelivu thanni*. Perceptions, beliefs and practices seem to differ from village to village.

A Caspian tern quarters by the beach, eyeing the jumping nets and flying sardines.

Tadpole Tank
27 December 2021

The ghost skins of granite ghost dragonflies perch on the walls and dangle from webs within the crumbling cement tank. And those of common blue-tails clutch on bloated alternanthera leaves. Eerie white echoes, which still have a grip. Holding a hollow shape of an earlier life. In the corners long-jawed orb-weavers have spread their horizontal webs. Littered with dead midges, mosquitoes and mayflies around a hairline which now and then flares into spider.

Mist has set in after the monsoon tapered. Mornings and nights are chilly and thick inside the lungs. Around 5.30 a.m. I hear the thick-knee calling from the empty roads, the first to cleave the heavy silence. The mosque sounds the salah soon after.

This was once vast marshland—part of the Pallikaranai wetland complex—which is within the Kovalam/Pallavaram watershed, the northernmost part of the Palar's basin overlapping with the southern reaches of the Adyar's basin. Water naturally converges here, rests and percolates at the season's end. It is its sleeping space—now largely turned to real estate. Water doesn't care, it claims it anyway.

There is a time after the rains when wetland surfaces begin to thicken and turn eutrophic with algae and duckweed. It has begun now. Life starts to congregate and crowd at this interface. It has become solid ground for fishing spiders, nursery web spiders and numerous froglets of ornamented pygmy frogs and cricket frogs. It is liquid enough for water striders and feeding tadpoles. I see a scouting party of carpenter ants crossing the floating duckweed and a grey wall-jumping spider roaming this seasonal surface. A German cockroach too. Land slowly crawls over this edge.

A kingfisher flies in to pick something off the surface and narrowly misses the jaws of a 2-foot stinging catfish. It thrashes out, aggressive and sudden, like the water's subconscious, and splashes back in. Duckweed covers its tracks. Maybe catfish wait and lurk under the surface-dwelling insects till a shadow flies in to fetch them. A startled moorhen throws a loud tantrum. I worry about the barn swallows sailing along inches from the ripples.

The cattail reeds in the marshes are seeding. Their fluff is carried by tricoloured munias to make soft linings for their nests.

Plain prinia fledglings hang around playfully among the reeds, five or six together, fat and red-beaked. Soft, airy, clear-noted pied bushchat song permeates the late morning.

The drying clay by the sidewalks wears the passage of snakes, crabs, egrets and openbill storks. A pair of blue river damselflies (*Pseudagrion microcephalum*) are mating on a blade of bulbostylis grass, their tails forming a perfect heart—the male summer-sky blue and the female algal green. Their joint flight, tail to neck and tail to body, puts me in a trance.

Velachery Marshland
2 January 2022

New-moon day. A nymph of a damselfly swims to a floating leaf and holds on to it with two forelegs. Its caudal gills wave behind. It is about to break out of its old skin, maybe now or two hours later. Two damselflies float, curled dead. The first day of the new year had seen sudden heavy rains in Chennai, lasting from noon till midnight. And these look like insects which drowned in the downpour at the time of eclosing.

December 2021 seems to have recast my emotional and philosophical relationship with death. It is a new entity now. Death has come visiting too often in the last two years. Thrice it has toppled my life. But death is a wise advisor and our eternal companion, says Don Juan the shaman to Carlos Castaneda, always at an arm's distance. 'Whenever you feel ... that everything is going wrong ... turn to your death and ask if that is so. Your death will tell you're wrong; that nothing matters outside its touch.'[2] I've written myself a death meditation—drawing from Krishnamurti, Maria Popova and Don Juan. From December-end

[2] Carlos Castaneda, *Journey to Ixtlan: Lessons of Don Juan* (Simon and Schuster, 1972), p. 33.

I've been reading it in the nights often, before planning the next day and noting down priorities. If I had another year, another week, another day to live, how would I do it? What would I give my energy to? I consult death like a friend, like the last Peverell brother in *The Tales of Beedle the Bard*. Its power of inevitability lends extreme clarity. Petty worries often drop, surface urgencies and lethargies disappear. Death shows what matters. Few others can tell you your core purpose right now other than death. I imagine it as a tall, hooded figure with a kind face and wisdom mortality cannot comprehend.

The webs of long-jawed orb-weavers hold drops of the morning's mist within the cement tank. I think about how much I understand living beings and connections between them entirely through the meaning 'webs' offer. Can we at all perceive interrelationships of life around us without the portals of metaphor spiders build for us? On the ground night train millipedes rummage the drenched leaf litter, turning death into life.

Dozens of sea pens have been turning up this week in the bottom-set nets of Urur Kuppam Beach. The black ink on their frills is, however, from snared cuttlefish. It's my first time seeing sea pens on the Chennai coast, another kind of creature-complex polyps create in the ocean, a puzzle between singular and plural, like anemone, man-o-war, sea whips and blue buttons. Sea pens lodge themselves into the mud and spread their feathery organs like living quills against a steady flow of plankton. They can uproot themselves and relocate when the currents change.

Another unusual and frequent bycatch this month is the Sipuncula, aka the peanut worm, which is a phylum by itself and one of the most distantly related being to every other animal around. I told Palayam Anna about them and dug his knowledge a bit. He said he finds them both on clayey nearshore beds during November–December when the *vadai kaathu* (wind blowing from the north) is strong and the olni current is present. They are seldom seen during the rest of the year in these parts.

In almost every story told by an artisanal fisher, the wind plays a prominent character. It is the mastermind and fate-maker. Every event is connected somehow to its strength, direction or just presence. I am often astounded by their 'anemo-memory'. Palayam Anna told me a story of how he once caught a large sailfish two decades ago, from when he set out in the morning to when he hooked it and lugged it into his boat after hours of pursuit, and his every sentence and stretch of hand summoned the wind as it was and as it turned on that day.

Detritivore Meditation
Velachery Marshland

It rained for a long time at night. The dry leaves of rain trees are piled up on the side of the street in a layered gel where the tarmac meets the cement pavement. Night train millipedes (*Orthomorpha coarcata*) have gathered on these patches, nibbling on the leaf cakes. They are called *train poochi* in Tamil. The full-grown ones are black with flame-yellow dots where their legs meet their long torsos. Their few hundred legs move and a single, slow ripple travels across their bodies.

Night train is such a poetic name. This morning they let me ride back to another rainy morning about a decade ago when I was watching millipedes. I was in my late teens. I had left home trying to run away from a six-year period of darkness. For years I feared sleeping because of the recurring nightmares. I developed a practice of seeking wisdom across species during inward storms. I'd watch a banyan, skylark or millipede and ask, 'Can you help me, please?' Then I'd hold my mind as a receptacle and pay full attention to the being, its energy, its living and my watching of it. And I let my intellect sink.

There would then be a spontaneous insight arising from this attentiveness, as if passed on from consciousness to consciousness across beings. This would be a moment of healing or meaning-making, helping me recast pain or confusion into constructive energy and wisdom.

I remember today the quiet teachings offered by millipedes to me many years ago. I unconsciously bring my hands together in gratitude or maybe in prayer. I let frogsong wash my mind for a while. Then I come back home and write a detritivore meditation.

(Detritivore meditation. To be read out slowly, with sufficient pauses between lines and longer pauses between paragraphs. To be done after rains where there is the smell of petrichor, preferably in the presence of millipedes or other detritivores.)

Gently close your eyes and bring your attention to your breath. Bring your attention to the sensation of air flowing through your nostrils, whether it is chill, warm or mild. Bring your attention to the smell of wet earth. Notice the feelings and sensations created by the awakening soil and the work of numerous beings in it. They are turning the past season's detritus into fresh, fertile soil as you breathe. Let your awareness stay with this smell for a while and the response it evokes in you.

Now, as the earth's smell enters your breath and bloodstream, look back at the events of your life and bring to mind a trauma which still affects you or a period which remains raw and unhealed. Bring it back as if it were happening now and give it the full intensity of your attention. Notice the sensations and feelings your body radiates as you experience this episode now. Enter its discomfort fully without avoiding it or turning away. Shine your awareness like a beam of light on this trauma, this detritus of the past.

As you pay attention to this experience, breathe deeply the wetness of the transforming earth. Invoke in your mind the millipede, which turns detritus into fertile earth, which heals soil for its own growth and that of other beings, including plants and trees. Invoke the millipede and its powers in your mind like a spirit creature—its form and its role (or any other detritivorous being abundant near you: earthworms, pillbugs, domino roaches, woodlice, etc.). Let the millipede's energy guide you to compost your trauma into something fertile and constructive, for yourself and for others in your life.

Bring back to mind again the event or period you still carry heavily inside you. Give it your full attention again. To the thoughts and images it brings, the feelings it creates throughout your body. Enter this fully and attend to it without turning away, letting it pass through your gut. Know that when you fully observe your pain it can transform into wisdom and personal power. Let the millipede's energy help you do this. As you observe yourself, notice how the intensity of your pain and its makeup in your mind begins to change. Know that when you compost your own suffering, you end the long chain of trauma possibly passed on through generations and many branches of people to you. When you compost your own suffering, you create peace and healing for others too.

Now bring to mind how this unhealed trauma, the fear or anger or anxiety associated with it, has guided your life. What decisions have you made or not made because of it? What habits and behaviours have you developed or not developed because of it? As you fully shine the light of your attention on it now, as your pain begins to transform into insight and wisdom, how will you let it guide you now? How will its location in your mind change? What actions would you like to take now? What decisions would you like to make now? Let the millipede guide you into transforming your suffering into your own unique, deep wisdom.

When you are done, bring your attention back to the smell of petrichor. Breathe in the soil's smell, the labour of numerous living beings deeply, and let it fill you. Gently open your eyes.

Guindy National Park
13 January 2022

Gossamer dawn. January mist fine-prints the vegetation, showing the land is rigged with spider webs. There might be miles and miles of webs within each square metre in front of the railway station. The grass is packed with funnel and sheet webs. Orb webs hang heavy across plant stems, branches, open sewers and fencing posts. Dew collects on dead carapaces parcelled on them. Long-jawed orb-weavers hitch their horizontal webs across marsh reeds and the three-dimensional trap nets of tent-web spiders are set up within the forking branches of young canthium shrubs.

A few days ago I tagged along with Rohith and Nanditha on their butterfly survey at Guindy National Park. I was instead snared by the forest's spiders. Spiky orb-weaver webs were in their hundreds, larger than umbrellas, tethered to tree branch on top and shrub below. They, like butterflies, prefer the edges where forest flows into fire lane and cart track. A spiky orb-weaver weaves distinct zones on her web. The 'hub' at the centre, her sensory HQ sewn with thick, wavy webbing, then a 'free zone' where just a few threads of web radially stretch, pure vibration carriers, leaving space between her resting space and dining mesh. The outermost zone is her 'capture area', her death trap, which she threads together in a 'sticky spiral'. She makes the toughest biomaterial on earth, four to five times as strong as steel though only a fraction of its weight. A sheet of her silk of pencil-width thickness can rebound a jet plane. She has a whole inventory of silk types she produces from seven different glands—swathing silk, egg-sac silk, non-stick dragline silk, anchoring threads, and inner and outer fibres of sticky silk which she spins around her capture area.

The egg sacs of another spider species, which builds cuboidal tangles between grasses, were hatching in multiple places—spewing plumes of scattering grey pixels, thousands of tiny spiderlings. They were simultaneously hatching across the forest.

The female spider exhibits such astonishing craftsmanship. To ask if she is 'intelligent' usually means we are stuck within our perceptive boundaries. What we often unwarily mean in our queries about thought, intelligence and consciousness of other life is how much of it is confined in a small space in the body, in the moment of time we choose to look, and how much of it resembles our own. Often what we mean by intelligence is similarity to ourselves. Our blindspots we consider the unintelligence of the other. But like Anansi of Asante folklore, spiders break such boundary conditions and perhaps can lend us that capacity too, like Anansi's stories did for the Jamaican plantation workers to rebel against plantation owners.

A spider has 'extended cognition', write scientists Japyassu and Laland. A spider weaves some of its own mental apparatus into its web, leaving some of its cognition stuck onto it.[3] A web is more living, sensing and thinking than we can tell. Spider and web together form a larger cognitive system. Webs process and conduct vibrational information, store memories through their differences of tension. And webs actively spring towards prey due to the electrostatic glue the maker leaves on them. They charge the air immediately around, possibly constantly giving the spider atmospheric information, tripping its field.

I think about Giorgio Parisi's Nobel-winning work—another marvellous discovery of systemic properties, 'extended cognition' in a sense, which functions in the same way in salt crystals as in starling murmurations. I ponder the imaginative influence spiders and their webs have had on human thinking. Could we have been able to abstract, extend perception and see the entangledness of the living world at all without the structural metaphors they weave for us? This web-like interconnectedness of everything, and the language we use to speak of it? Could it possibly be that this element of our imagination/perception is also part of their extended cognition?

[3] H.F. Japyassu and K.N. Laland, 'Extended Spider Cognition', *Animal Cognition*, May 2017.

Neelankarai Beach
20 January 2022

Millions of blue buttons, sea swallows and other floating creatures have washed ashore on Neelankarai Beach. Kilometres of strandlines are made of the small, shrivelled sea creatures. I walk along their disintegrating remains and inspect the different species. I call up and check with other friends and confirm that they are being seen all along Chennai's coast.

What is common to all these creatures? They make up the neustonic community in this ocean. All of them live at the surface of the water. A world of beings that have adapted to live on this thin film, this meagre stratum, which is totally different from the water below or the air above in its physicochemical properties. What has dumped them all on the shore?

What brought me first into regularly walking the coast was the blue button (called *netti* in Tamil). In January 2019, I was out surveying shorebirds at the Adyar Estuary with a group of birders. It was low tide at the time and a similar mass beaching of blue buttons had happened. I left the birds and crouched along the intertidal. I held them unknowingly and after a while their tentacles stung mildly, like salt on a bruise. A blue button is not a single creature. Each seemingly dime-sized organism is a free-floating carnivorous creature-village made of polyps. Polyps themselves are small, jellyfish-like beings. These too are a puzzle between singular and plural, queer creatures, beyond the scope of our pronouns.

The blue button's physiology is communal. On it some polyps group up to play the tentacles, stinging and reeling in prey. Some make up the oral chamber beneath, through which if you look closely, you see some remains of tiny crustaceans. Some do the digestion and distribution of food. And some harden up and create a float at the centre which holds up the blue button to the sea surface. There are rings on the float which grow just like and resemble tree rings. A ridge for monsoon and a furrow for rainless winter or harsh summer, perhaps. The blue button's back is possibly a calendar of seasons, as felt in the open seas.

Floating blue button schools in the open sea also host a larger community specifically adapted to live among and feed on them. The sea swallow (*Glaucus sp.*) is a blue and white sea slug which looks like a piece of sheet lightning floating in the water. It catches and climbs on surface tension like it were monkey bars. It eats blue buttons, stores their nematocysts (stinging cells) and acquires their ability to sting. Sea slugs in other habitats too show the ability of absorption from their food. Back on Kovalam Beach where ulva seaweed grows on the edges of groyne rocks, you sometimes see—if you look arduously close—the elysia slug. It is small, flat and green like a rogue piece of seaweed. It eats these plants, stores their chloroplasts and begins to photosynthesize.

Lying among the masses of sea swallows in Neelankarai I find a couple of purple storm snails (*Janthina sp.*). They blow a bubble balloon which keeps them afloat by the surface near their blue button prey. Then there are by-the-wind sailors—a living sailboat with tentacles below and sail above to be wind-steered on the waters. Another polyp-made marine pilgrim. There are also hordes of goose barnacle lattices sticking on thermocol bits, table tennis balls, drift footwear, drift ballpoint pens and liquor bottles.

Fish like trevallies take refuge under the floating blue button schools when young, and the masses of stinging tentacles give them protection from predators. Experiments have been done where the fish and blue buttons are separated, but when put together again each fish will return to its respective partner button. Young fish form relationships with specific buttons and hang around under them. Fisherfolk say that the pomfret (*vavval meen* in Tamil) when young spends time under orange sea nettle (*manja sori* in Tamil) schools and that it's their *thai veedu* (maternal home). They say that the jellyfish don't harm them alone, but that is possibly because the pomfret young have protective mucus lining.

These surface communities are bound to increase in their numbers and beachings with sea surface temperatures rising, making waters conducive for proliferating jellyfish and jellyfish-like creatures. Though some young fish take refuge under them, large amounts of jellyfish—ocean jellyfication—threaten fish populations. In some of the shore seines cast by Neelankarai fishers in the past few weeks, anchovies and sardines came mixed, stung and mutilated with large numbers of sea nettle, and it was impossible to separate them. About four hours of work by more than fifty men goes to waste when sea nettle groups float nearshore.

For the first fifteen days of January, winds have been shoreward and strong from the Bay of Bengal. Has the downwelling it created cast out these creatures en masse from the ocean? In December 2020 just after Cyclone Nivar we saw blue buttons in Chennai, and larger strandings were reported in other parts of the Tamil Nadu coast. Sometimes they strand after a storm. Sometimes they strand inscrutably, to keep us guessing why. Far away on Australian beaches too man-o-wars are stranding en masse all through this week, and an earthquake near New Zealand is thought to have caused it. Maybe the quake's wave energy travelled to this coast, or maybe these creatures indicate seismic activity from somewhere else.

Between June and August 2019 marine activist and distance swimmer Ben Lecomte was swimming through the Great Pacific Garbage Patch.[4] There are large garbage patches in all five ocean gyres—the North Atlantic Gyre, the South Atlantic Gyre, the North Pacific Gyre, the South Pacific Gyre and the Indian

[4] Amanda Heidt, 'The Constellation of Creatures Inhabiting the Ocean Surface', *Scientist*, January 2023.

Ocean Gyre. These are places where surface currents meet and concentrate all the garbage released by humans into the coastal waters of the earth's oceans. Ben, and the marine scientists he was helping, found that as they neared the garbage patch the blue button neustonic community rapidly increased in density. There was as much blue button and associated life here as debris in each cupful. Pelagic birds and sea creatures ate plastic which looked like the blue button. It is possible that once upon a time, only this blue floating community lived densely in the gyre. This was their stronghold, till polymer trash began to gather. When currents deviated or were disturbed for different reasons maybe some blue button communities were pulled from the gyres and beached on the shores. Now they always beach along with floating garbage. I am certain that if we check the Indian Ocean Gyre—another garbage patch—we would find a thriving blue button refuge amidst deadly plastic squalor.

Today the wind is absent. Longshore currents are absent. Stubby waves roll in slowly, one by one and far apart. The sea is sleepy and silent. A gull flock lazily bobs on the water 10 metres from the low-tide line. Neelankarai Kuppam's fisherfolk return from the Thiruvanmiyur kallu. Their nets bring large rabbitfish (*ora meen*), butterfly rays (*kuruvi thirukkai*) and torpedo rays (*thimilai*). Two men dance as their hands untangle them, as the rays expel their last electric bursts. I try to wash off the sand from a torpedo for a picture and get a small shock. I remember Palayam Anna's story about giant devil rays (*kottan thirukkai*). He said they were terrible creatures and that he hated them. I told him how they were harmless plankton feeders which swim gracefully like large flying eagles. He said, 'All that is fine. But have you seen a bear rubbing its back on a tree? Like a bear, the devil ray will come to a boat or the anchor rope and give itself a nice scratch. We sit there being tossed from one end to another. I have been toppled over once.'

It is overwhelming, the stories the ocean keeps bringing. Day by day it becomes vaster, more wonderful and mysterious. These are important experiences we need as human beings, psychologically, spiritually, collectively. While accepting the John Burroughs medal in 1952, Rachel Carson prophetically said, 'Wonder and humility are wholesome emotions and they do not exist side by side with a lust for destruction.'

With my friends and fellow nature educators at the Palluyir Trust, I have begun what is to be a year-long coastal curriculum for children of the fisher community at Urur Kuppam and Olcott Kuppam, in collaboration with Prem, a friend and phenomenal teacher from the community. Their home beach near the Adyar Estuary is the primary learning space. It was our first session today, about finding and grouping coastal biodiversity on the dockyard, and beached blue buttons graced our class.

Perumbakkam/Sholinganallur Marsh
28 January 2022

Wide as a housing plot, a wedge of large waterbirds swims towards Pallikaranai in the low sky. It is windy and cold near the marshland. The nose and shorter arm of the V-shaped flock are made of three painted storks, and the longer arm of six pelicans. The group glides and beats their wings in slow pulses, which ripple from stork no. 1, one by one, to pelican no. 6.

On the ground a skylark with its crest open forages in the seeding grasses amongst yellow wagtails and little ringed plovers. Two or more larks are high up somewhere, higher than the barn swallows, only heard deep from the clouds as rain hits my forehead. In the back of my mind Bharathidasan's poem '*Vaanampaadi*' runs on loop. '*Vaanamdhan paaditraa? Vaanilavu paaditraa? Thenai arundhi siruthumbi meleri nallisai nalgitraa? Nadungum idikuralum mellisai payinru miga inimai thandhadhuvo? ... Aendhum vaan vellatthil inbavellam thaankalakka neendhukinra vaanampaadikku nighazhthinen.*' (Is the sky singing? Is the moon singing? Did a dragonfly drink nectar, climb into the clouds and make music? Has thunder suddenly decided to turn melodious? ... In the flood of the sky, mixing its flood of joy, I stood watching the skylark swimming, singing.)

Thousands of ducks have come to the waters of the Perumbakkam/Sholinganallur marsh from the Holarctic. This is their winter home. Marshes in these months are striking for the multispecies communities they hold. Many grassy mud islets are shared by birds of different feathers, 'murmurations of difference'—as ornithologist Drew Lanham described in avian terms the coming together of diverse people.[5]

Pelicans, painted storks, pintails, stilts, grey herons, cormorants and bee-eaters all colourfully mixed, sitting, basking, sleeping, preening. I stand and watch, evolutionarily distant from such a cross-species coming together. Something that is so difficult oftentimes between different groups of the same species, as in my own. A swamphen runs spread-winged at a purple heron staring into the water ipomoea and takes its spot. A soaring harrier sends a storm of ducks across the marsh and into the wetlands enclosed within the Elcott special economic zone, caught between large IT buildings. Glossy ibis and jacana forage side by side on the water hyacinth.

I ask my friend Karthika Chandran, a researcher of mixed-species flocks among birds in the Western Ghats,[6] what she has learnt from such cross-species coming-togethers. She says, 'I see these birds choosing their associates (or "friends" if I could say) in mixed flocks very smartly—they usually come together with those species that bring out the best in them or make their lives a

[5] Drew Lanham, 'A Convergent Imagining', *Emergence Magazine*, February 2021.

[6] Karthika Chandran and Vishnudas C.K., 'A Comparative Study of Mixed-Species Bird Flocks in Shaded Coffee Plantation and Natural Forest in Wayanad, Kerala', *Indian Birds*, October 2018.

little easier. And that is something I feel is very important for any of us to have—a good friend! Also, I was always keen to watch how the birds shared their little spaces with each other, trying to adjust most of the time, and yet ever-ready to fight for what they really want, if anyone goes overboard. So, choosing friends wisely and being accommodative of others' needs while still having boundaries were some of the many lessons I have learnt from them.'

Today is too cold and cloudy for most birds in the marsh. They all remain floating, sluggish or crouched on the islets. On sunnier mornings hunting fish is another common cross-species collaboration. A block of cormorants takes the frontline, swims in a rhythm of breaching necks, pursuing fish and flying forward like a conveyor belt. Rows of pelicans paddle right behind, plunging their pouches to scoop gallons of water and filtering fish like baleen whales. Herons or storks may stride by the side, picking up escapees. A flock of whiskered and Caspian terns follow the swimming birds, plummeting upon leaping tilapia, mullet and catla.

Over my home now openbill storks fly dispersed through the florid sky every dawn from around 6.10 a.m. to 7 a.m., southwest towards Pallikaranai, filling the whole firmament. Then at dusk they glide northeast again, around 5.30 p.m., back to their roost somewhere in the city.

Tadpole Tank
30 January 2022

During the maturing of cricket frog eggs, in the first three days, the pure yellow protoplasm inside condenses into tadpole embryos. The egg walls bloat and become barely visible. The clutch tethered gelatinously to an aquatic stem drops in density, rises and spreads on the water's surface. They are a patch on the water, like a hundred tiny commas suspended together, unmoving even when ripples move through it, like a close-by constellation. The embryos are smaller than pencil nibs. They'll hatch in three days and swim around as clear glass orbs propelled by hair-thin black tails.

You have to hold your eyes 20 centimetres or less from the clutch to notice the occasional turns and tail flicks of the living shapes within the eggs. There is a courting corner in the tank, the shadiest spot, which adult cricket frogs use for their amplexus and where most of the egg clutches are found. Within the tadpole tank, pond-skaters perch on the embryo films, just like they do on floating algae, to rest. I don't see them prey on the frog babies.

A long-jawed orb-weaver somehow has fallen off its horizontal web into the water and is grabbed by skaters. Three fight over and snatch the spider-meal from one another. Water is trampoline for the water-skater. Whatever it does, the meniscus never breaks to its movements, but folds and rebounds. Its hairy hydrophobic legs crease the fluid plane and the insect kicks back, rowing with its long middle pair. I see a mating couple—the larger female carrying the male around as it picks up floating dead mayflies and feeds itself with its forelegs. Moving shadows startle her at one point and she jumps nearly a foot into the air, landing with the male still piggyback. It is as though water is a sheet of rubber for them. Skater females also lay their little oval eggs in the cricket frog courting corner, on the wall or on the submerged stems. Red eyes develop on them soon, all facing the sky.

Froglets swimming in the water with their newly grown forelimbs can't wait to venture out of the dilapidated cement walls. They swim up from the algal jungle at the bottom, leap out like little dolphins and plop back in.

I have seen on a few occasions, once when it was drizzling, adult skaters herd their little ones into a corner of the tank and slide around to keep them inside a skater crèche. It looks like parental care to me. For the most part, young skaters dart around on their own and spend much time co-feeding with groups of velvet water bugs, on the range of life which dies, falls on or floats up to dent the water's veneer. Water bugs are more minuscule than mustard seeds, and they spin, sprint, crowd and carrom over food like a volatile chemical reaction—like elemental sodium exposed to air—but not so much as causing a dimple on the water for all their frenzy. They feast on freshly laid frog eggs.

On some days I am lost for a good two hours gazing into the tadpole tank, bending over or squatting over its walls, taking pictures. Azhagamma, who handles the motor room and the watering of plants near the railway station, yells at me almost every day, as she takes me for one of the drunken men who use the undrinkable algae-filled water to dilute their alcohol, and who leave their trash and bottles around. I have explained my intentions, showed her pictures and befriended her, but her eyesight is not very good so she still comes screaming expletives at me from a distance till she can recognize my face.

Common tree-frog tadpoles eat algae like dugongs eat seagrass. Their mouths are underneath where their chins should be and they browse the floor and wall bottoms. If they want a morsel floating on top they have to wriggle up, flip upside down to grab it and wriggle back down. Pygmy frog tadpoles have copper-coloured spines and their mouths are on their foreheads. So they feed like a school of sardines gulping at the surface, making the water shimmer. There is possibly a symbiosis between the skaters and tadpoles. They often hang around, above and below, together. The tadpoles are closely observant of skater movements, and if they notice the insects dash in panic from an aerial threat, they dart deeper into water. And I think skaters watch tadpole activity too, which might signal aquatic threats to them like predaceous fish.

On a sunken chunk of coconut fibre, the outer skin of a granite ghost's nymph is ripping off as it prowls around. I feel a second-hand thrill for the nymph, who is very soon going to climb out and fly for the first time as a dragonfly.

Kotturpuram Urban Forest
8 February 2022

There are stacked galleries of tent-web spiders (*Cyrtophora cicatrosa*) in the bamboo clumps near the northern compound of Kotturpuram Urban Forest. And at least half a dozen or more of their webs, these feats of craftsmanship, on every Ceylon ironwood tree (*Manilkara hexandra*). Sunlight is scattered blue by atmosphere. Here, today, late-morning light is scattered sharp violet by overhead tent webs.

Seven friends are painting scenes of biodiversity on the rock slabs in the urban forest. And I give myself three hours of 'watching' spiders, ants and other creatures as the artists do their work. I sit before tent-web spiders for the most time, with a pen and notebook. I think of whether spiders feel bored in their almost all-day alert stillness. And if a state of boredom ever exists for a creature of such profound solitude. Would the world still be recognizable to me—this forest, the Adyar River flowing to its north, the passing clouds above—through its eight eyes. I am practising staying in and feeling comfortable in the web of my own mind for periods of time, without trying to escape its strangenesses and vicissitudes. Its architecture reflects not the simple radial pattern of orb webs, but the inscrutable three-dimensional tangles of a tent web. When I watch spiders for long enough, I reel into thinking about the mind.

———

Line break ...
a spider hurries
over the words.

Boredom, I am learning and noticing, is essential for my writing practice. It is the phase when the mind sensitizes, craves stimulation but also recovers from it. Thirty minutes or more into boredom, if allowed, a portal of perception opens which is moved by the little things, notices the subtleties of shades on sand, the variability of sea waves, the trills in the wagtail's song, the feeling of dew on bare toes. Psychiatrist Anna Lembke talks about embracing boredom—going on a dopamine fast, to let the brain heal from too much pleasure and the intense craving that follows, narrowing our capacity to experience any joy as this cycle intensifies—a state she calls 'anhedonia'.[7]

I have reframed the meaning of boredom for myself and am beginning to embrace it as a fertile discomfort. The modern world's digital desensitization and overstimulation numbs our minds, glazes our eyes and we risk no longer being able to experience wonderment at the spider's web, the wasp digging its burrow, or the smell of wet earth. And I cannot live like that; I would lose all meaning and purpose. Boredom could be the body flagging that a world of ordinary—yet extraordinary—sensory experience is present unfelt. Other neuroscientists like Alicia Walf and Nicholas Kardaras show through their work how essential boredom is for creativity, insight, resourcefulness and empathy, especially in the brain development of children.[8] And not in the least, learning to stay with boredom also beats the algorithms of advertisers working to monopolize our attention. So I allow myself to be bored, feel that fidgetiness and fecund sensory aching, till a threshold is crossed in the mind. Then the edges take centre, the ordinary acquires extraordinary aspects.

---

[7] Anna Lembke, *Dopamine Nation* (Dutton, 2021).

[8] Neurosciencenews.com, 'Let Your Brain Rest: Boredom Can Be Good for Your Health', 30 August 2020; Nicholas Kardaras, *Glow Kids: How Screen Addiction Is Hijacking Our Kids—And How to Break the Trance* (St. Martin's Press, 2016).

Tent-web spiders make webs of unique design. They build a large volume of tangle webs with two horizontal nets cutting across its centre, with a mesh so fine and consistent across their spread that they could have been measured and laser-made. The top net is anchored like a circus tent at the cone of which the spider rests upside down. Its green egg sacs may hang like a plumbline above it, along the perfect structural centre. The tent is ripped up a bit at its apex, for the spider to burst into the tangles if a moth or fly is caught. Its tangle fort is thought to let entry of its prey but filter out anything predator-sized. In the bamboo clump, however, the webs form a massive matrix. The ends of different tent-webs merge and scaffold onto each other, the females on the extremities anchoring the structure to the stems, forming truly a community fortress.

A signature spider has made its web 3 feet from and angled towards a paper wasp nest under a pandanus plant and is wrapping up a wasp it has caught. Dewdrop spiders (*Argyrodes sp.*)—little glowing blue beads—hang around at the edges. They are creatures which wait on orb webs and feed on morsels too small for the larger spiders or on leftovers. Lines of Jo Mariner's poem about J.A. Baker's intense, enduring observational treatise *The Peregrine* run in my mind: 'love// meaning live in intimate observation ... wanting others to know/wanting it purely ...'

Yes. Love ... meaning, to live in intimate observation.

Apartment Terrace, Velachery Marshland
12 February 2022

5.50 a.m.: The dark of the eastern sky begins to ferment.

6.10 a.m.: First call of the koel. Sudden and alarmed, as if it has overslept and been startled awake, late for work. Cricket stridulation stops like a switch. Pied wagtail begins to sing. Francolin and brown shrike start to call.

6.20 a.m.: The sky has cloud belts of rosemilk pink. Then it turns LCD white. Mist-muffled crow pheasant hoots sound like pelican wing beats.

What fickle things dawn and dusk have been this month. Sometimes stunning, other times simply a snuffing out. Sky is a strange creature. Sky is also a form of self-love.

Being present at twilight lets me think from a liminality, from an intertidal space within. I've read that it's also a prerequisite for our brain's clock—called the suprachiasmatic nucleus.[9] Our eyes do so much more than image-forming. There is a circuit in our retina which exists purely to see sunrises and sunsets. It tells the brain their contrast of colours, tunes its clock to their rhythms which then signals the function of every cell in the body. We are like flowers, we are photoperiodic, constantly unconsciously scanning light. Dawn and dusk hours tell us when to bloom open and when to close. Tear down those skyscrapers. Twilight, more than any other light, penetrates our bodies and minds. I also notice daily a certain peace and poetry of thought, converging only at this time. So much life comes alive during this hour, so vocally, almost in ritual acknowledgement.

6.30 p.m.: Dusk glows brick red, and goes off like a flame. Pipistrelle bats forage in the overhead mosquito columns.

6.40 p.m.: Openbill storks fly back from the marshland. Flying foxes take over the orange sky.

6.50 p.m.: Fifteen or so little swifts are very active, chasing each other and chittering over the railway towers. The western sky is bands of blood-clot blue. Then dying embers. Then a grape-skin glow, even till after 7.

[9] M.U. Gillette and S.A. Tischkau, 'Suprachiasmatic Nucleus: The Brain's Circadian Clock', *Recent Progress in Hormone Research*, 1999.

Section 5

DIFFERENCE
MEDITATION

Tadpole Tank
11 March 2022

Life invoked by the rain lives long after the last showers. Even deep into summer.

At 7 a.m. I come to the tadpole tank to see a tangle of legs and knotting of water. A dragonfly has fallen in, possibly when it was eclosing. There is new exuvia nearby, clutching a cement block. The insect's wings are still crumpled, its body still curled. More than a dozen pond-skaters are gorging on it. They shove each other around, jump over one another, block each other's spaces. Some of their back legs lift off the water.

Rippled sky—
the creases made
by water-skater legs.

———

There is now just an inch of water in this tank and its continuing busyness of life is amazing. Three weeks ago I planned a field session for the next batch of UWW interns—starting at the tadpole tank, then to the growths of thumbai and candlebush by the train station, onward to a stand of neem trees showing heat stress and dieback, closing with two patches of marshland to see some waterbirds. But the group stayed engrossed with the creatures in the tank for an hour, doing a pond survey sheet, and the walk time got over soon after that. This morning I showed tadpoles and water-skaters to boys from the nearby government school, and the dragonfly being devoured. Last month they asked me, 'What is the point of watching all this?' Now they simply join me, point things out, ask for names. One tells me he saw a flapshell turtle in this tank last monsoon.

Hidden in the algal forest at the tank's bottom, I noticed little, crocodile-like creatures some days ago. Tadpoles of a different frog, revealed now by receding water. I called them croc-poles until I found out what they were. They lie over one another like pebbles under a stream, flattened above and flattened below. They nibble on algal tufts underneath—their home is their staple food. But mysteriously, every couple of minutes or so, they dash to the surface, touch it with their mouths (located on their chins) and wriggle back below. This scares the water-skaters and sends them jumping away at times, making the swimming pygmy frog tadpoles and backswimmer insects dart off. I thought maybe the croc-poles are omnivorous; or maybe they are just darting to gulp some tasty debris, and the skaters and swimmers are moving out of the way to avoid being whacked. Perhaps as the tadpoles grow, and their breathing shifts from gills to skin and nostrils, they swim up to take gulps of air now and then. I don't know. A 5x3 feet body of seasonal water gets deeper and deeper the more one observes. There is this magic whose workings I am trying to understand, of simple daily observation, which can turn a pond or person into a cosmos of their own. To repeat Nan Shepherd's words, 'The thing to be known grows with the knowing.' Attention is the most magical substance in the universe.

The croc-poles soon develop bars on their thighs and warts on their back, turning into cricket frogs. I watch freshly metamorphosed frogs hunt backswimmers today. One clutches the cement wall and takes its time to scope the insect. Then it flings itself into the water and comes out to perch on matted algae, the insect already swallowed. Another on a brick drags a swimmer by its rowing legs, snaps down on it four times and swallows it.

For the past three months, I've seen several batches of the fish-like surface-feeding tadpoles of pygmy frogs. But they never turn into frogs, not since the monsoon receded in January. They grow till they show the brown tuning-fork mark on their backs and their four legs, then they seem to vanish. Are the cricket frogs or the checkered keelback eating them? Is something stopping them from reaching full froghood? Are they all together hiding under a rock?

Also, where do the water-skaters go when the waters recede? And how do they return when the waters get higher?

A kingfisher feeds a younger one a knot of green algae picked up from the water's surface. Junior spits it out twice, not quite liking the taste. But the parent insists and finally the fledgling picks it up and gulps it down. Why do kingfishers eat algae?

Urur Kuppam Beach
30 March 2022

The edappu is the closest I've seen to the beach in a long time. Edappu is the local name for the junction of two waters, ocean colour fronts, a word which does not have an English equivalent to my knowledge. It usually refers to the zone where the muddy nearshore waters transition into the darker blue of the offshore. But the colours may show more subtleties, like today. There is chocolate-brown water for about 40–50 metres from the low-tide line. Then a narrow band of light turquoise, more green than blue. After that a vastness of dark blue, far darker than the sky. Bands of three distinct ocean colours of increasing widths, like a giant climate-stripes chart.

Ark shells are beached in plenty today, more than other bivalves. And so are several empty egg sacs of chank snails. The ocean doesn't merely reflect the sky. Its colours come of itself, though prominent only when the sun is subdued. Winds, plankton, algae, sediments, currents, bathymetry and mixing of layers give it its complexions, which for the keen eye is ever-varying day to day.

It is olive ridley sea turtle season all along the east coast. While the nesting is at its peak in Odisha, at Urur Kuppam Beach in Chennai the hatchlings of February nests are emerging inside the forest department's hatchery. Families, schoolchildren, government officers and half a dozen naturalists stand around to watch the baby turtles being set out to sea. They crawl determinately in the water's direction, mostly. Some walk sideways and are set in the right direction by the forest department staff and volunteers from the Students' Sea Turtle Conservation Network. Several hatchlings try to hug or climb another and piggyback their way to the waves. They try to swim with their small flippers as soon as the tide strikes them. Rolling waves throw them out a few times, and they crawl in again, while we stand planted in the sand so as to not stomp on them. They're eventually taken in by the receding water. The sand on the beach berm is a gorgeous etching of hundreds of tiny flippers.

Certain paths of thinking can happen only when the eyes can gaze far out, unimpeded, like over the expanse of the ocean. Certain branches of imagination remain blocked within confined walls and trapped vision. I come back in the evening wondering about 'natal homing'. How adult olive ridleys come back to their birth beaches to nest, for generations, being among the most site-fidel of sea turtles. What do they do if they find that their beach is eroded or filled with hard rock? I don't know. This affinity to 'come back home' in some sense is possibly there in all living creatures. In pelicans and spoonbills, as in mackerel, frogs and rock bees. And in us too, perhaps. I've been walking this stretch of beach since I was a toddler. It carries a sense of home and childhood for me, as it will for these baby turtles. Coming back 'home' could also be to an emotional landscape, a land of stories and memories which let me belong. Notice when we are sick or sad, what we return to, where we return to emotionally, behaviourally, inwardly. The strong instinct to return and relive. During our creaturely childhood, these events shape us cellularly, down to each nerve, then later beckon us to come back. When it is time to breed, frogs are known to seek out the odour of 'native ponds' where they spent their time as tadpoles. Just as sea turtles return and search for nesting places in single-use-plastic-laden beaches.

I think of how this is true for human beings. A return to wounded homes, emotions, memories. I have seen and read of many instances of parents replicating for their children the experiences they received in childhood. If this was harmful and even if they did not like it, they end up emulating it, as the events in childhood are known to enter deep into our psyches and craft its very structure. Parents replicate care or harm in the forms they faced because this is what home gave them and this is what home and kin have come to mean. This plays out subconsciously if there is no effort to consciously observe, intervene or seek help to resolve. Children, just as turtles and pelicans, are born in these wounded lands and waters, and invariably wounded homes. Yet, their childhood experiences and places make the world and life acquire meaning and purpose. In seeing this happen and acknowledging that we ourselves may be wounded in different ways, what becomes our work as educators and as human beings?

Railway Cement Pits
23 April 2022

The master understands that the pieces of the chariot are useless without the connectedness of the whole.
—Tao Te Ching

The tadpole tank has gone dry. So I now go to three square cement pits some 6 feet deep, where railway construction has been abandoned and water has collected. And with it, so much life.

In response to any aerial threat water-skaters congregate along one corner of a waterbody, packing themselves in a lattice. A friend told me they look similar to a cross-section of alpha quartz. When they come together in this fashion their nature changes. They stop their quarrelling—something they are always doing—and get into a crystal of interlocked legs. More aggregate to the sides of the existing crowd and fit into place.

Water-skaters resemble crystals, crystals resemble bird murmurations, murmurations mimic the shapes of other creatures. When matter or beings come together they are no longer the sum of their parts. There is an aliveness and intelligence the 'togetherness' gives rise to—that which makes the lake, the grassland, the estuary an organism on its own. Our laws and policies are not good at recognizing this. Rights and identities are ascribed to what is familiar to us as separate entities, who may themselves each be a community of selves. In human society, for instance, blame and crime is ascribed to what we define as the 'individual', without systemic responsibility.

The beingness of the whole wetland and the entire termite hill is a systemic property. But to see this, one has to be part of the lake or observe the anthill or even a group of humans face to face to a point where one's mind merges with its matter. The connectedness between things creates entities, more and differently alive and agent than its components. This sense of expansive selfhoods and compound beings is difficult to arrive at through thinking. It cannot be seen from the surface or the periphery. One needs to meet it disarmed, one has to experience it in the real world and see one's self in entanglement with the rest.

The last thing the human child starts to learn is entanglement because they've been shut off from the rest of the world for most of the time in a series of rooms. Deep, direct observation of others or of oneself is not a part of any syllabus. And their maturity is artificially delayed, so that their worldview can be set in a specific groove according to some ideology. It is through simple observations—such as watching water-skaters, for instance, or a tree or a file of ants—that larger questions arise and deepen one's seeing and experience of relationalities. How do we teach or allow for this to happen?

When an ant falls into the water, slipping off a dusty cement pillar, the nature of water-skaters transforms. In two minutes they crowd over the ant like wild dogs. Latecomers dash and kick the ones on the edges to create a gap. Those at the centre pivot and try to leave no space.

Coromandel dartlets like to lay their eggs on aquatic snails sitting on floating algae. Their mating pairs—thin, hovering stems—are at peak numbers now at the pits under the railway station. Most raft spiders scampering over the water are holding egg sacs or carrying backs full of babies. Water-skater females have laid eggs collectively on the western wall of the middle pit, just centimetres above the water. The insects actually climb the wall and lay their eggs, and sometimes stand guard over them, spreading their feet; not made for solid surfaces, like spiders. Yesterday I saw a female (most likely) guard the whole clutch with ferocity, charging at every insect which came nearby from all three directions. She even mobbed a raft spider on the wall, and the spider charged back at her. There were eggs of at least ten other females there. She was guarding them all.

The central practice of a naturalist is to observe keenly and, as much as possible, to observe beyond the threshold of one's comfort. I am learning how to be a naturalist in the wilderness of my body/mind. All observation in some way is an observation of this wilderness—how the body/mind responds to where it is, what it meets, its many creatures and weathers.

I see myself struggle a bit with water-skaters. Ideas deep and true in the culture of my species (or a part of it) make it challenging to understand distant kin. Unlike me, a skater sees matter as an undifferentiated cycling of energy. It simply eats anything living or was once alive which falls or floats on the water. Carpenter ants are the most common prey as they easily slip off the cement when scurrying sideways, gravity pulling on their flanks. But a dragonfly, damselfly, orb-weaver or mayfly will do just as well.

There is no concept of courtship among skaters. Each male chases a female, jumps on her and shakes her till mating ensues. It is said that such shaking makes the female susceptible to predators from above, so she agrees to mate if the male is persistent. No songs, no wooing, no letters or the like. Just jump and throttle. In the open water the insects incessantly mob each other, dash, chase, fight for space. Yet in the shaded parts they sit closely in a quiet lattice, touching leg tip to leg tip and sharing space.

I saw a hornet perch on the algae and drink water on an afternoon a few days ago. So I began keeping a plate of water for the paper wasps nesting on the vacuum cleaner in the balcony of my house. (I have been stung thrice trying to take pictures of their nest, which is slightly hidden from my view.) Many times a day the wasps sit and drink from the plate. Approaching summer has made black crazy ants (called *pillaiyar erumbu* in Tamil) move up from the ground floor and nest in the pots on the balcony of the second floor. Neems, copperpods, tabebuias and golden showers are in full bloom.

Leafcutter bees are cutting perfect semi-circles on the fresh foliage of the Indian beech trees along the compound of the apartment building.

Walking back from the water tank I see a juvenile drongo with a still-squarish tail and brownish feathers sitting on the branch of an Indian ash. An adult or parent comes to sit near it and junior flies away to the neem. The parent follows with an insect in its beak. The young one flies to a fence, screeching loudly, perhaps asking its parent to leave it alone. This continues for a while. The drongo summons back memories of my own behaviour in the past, evoking the similarity of our creatureliness.

Kotturpuram Urban Forest
26 April 2022

A parasitoid wasp gathering is taking place in the cluster fig's canopy at Kotturpuram Urban Forest. Hundreds of tiny wasps are perching under the lowermost fig leaves. Posing like yoga-doing scorpions, emerald and metallic apocrypta wasps are drilling into the unripe fruit with their zinc-tipped ovipositors and laying their eggs. The orange-coloured, more numerous sycophaga wasps are crowding at the fig's pores to enter and deposit eggs within. Neither species pollinates the fig, though. The tiny black ceratosolen pollinator wasp is extremely rare to sight on the fig trees within the city.

Summer noon—
wasps drink water
under the clotheslines.

The roadside peepals are putting on a spectacular show. They were all nearly bare last month. Their fallen leaves skeletonize to reveal every fine detail of their veination, and their branches gradually grow leaflets, red like watermelon and strawberry. They rustle stickily, silently. Their canopies stand out on all the streets if you see them from any terrace. The leaves soon begin to make chlorophyll which mixes with the red pigments, and for a few days the peepals are puke-brown. The leaves turn a dirty sludge-green as they grow larger, though they still feel like wet skin. Then they become a green brighter than acres of fresh paddy, gleaming in moonlight and streetlight like breaking waves at night. Most trees are now in this phase, laying down their cuticles as they turn darker green, reclaiming the torrential sound only peepals can make.

The gaudy sunsets and sunrises of red, gold, pink and purple are gone with the second week of April. In the thick of summer, twilight is quite plain, just gradually fading light. My mother and I miss the March sunsets on our evening walks.

On Sunday Nikkitha, Bianca and I did a module on butterflies for the children of Urur Olcott Kuppam. This is part of a yearlong nature-based apprenticeship we are doing for children of the fisher community. In this week's session we learnt to draw a tawny coster, looked at different butterfly families and life histories of common species and their habitats in Chennai.

Then we set off to Kotturpuram Urban Forest to observe butterflies in the field. April-end is not the best time for these insects but some children spotted and wrote descriptions of fourteen species. Among them the palmfly, tricoloured flat, blue Mormon, common jay and tailed jay were favourites. When they can understand through exploration and discovery, children's intrinsic motivation to learn and to feel wonder comes alive. Classroom learning is by and large extrinsically driven—by instruction, pressure, exam, punishment and so on. Outdoor learning can be intrinsically driven, tapping into an evolutionary need to explore, observe, share one's findings—as if we were wonder foragers. Many of these children are first-generation school-goers, whose education took a big beating during the lockdown. But I saw, when it came to butterfly names, the prowess they showed—in learning the spellings, writing and speaking the words right, putting down their observations and colouring, finding and sharing—was simply amazing. Even among those for whom language was a difficulty. I saw scientific inquiry emerge on its own in the questions they asked and translate into connections—butterflies they'd seen on their home beach, when, which and where. We debated a prevalent superstition the children narrated about the crimson rose butterfly—that if one caught it, smeared its red scales on one's forehead and buried it in the sand, money would appear in the burial site the next day.

Penny Whitehouse's advice comes to mind: 'Don't wait until your child's school understands how important green time is for their growing minds. Today, leave the homework untouched, in favour of outdoor play and real-world learning.'[1]

---

[1] Penny Whitehouse and Emma Bear, *The Muddy Chef* (Wild Dog, 2022).

Urur Kuppam Beach
27 April 2022

The white-throated kingfisher's is the first call at dawn these days. By 4.30 a.m. its rapid series of high-pitched notes echo from the train station towers.

At 5.50 a.m. the air is at a standstill at Urur Kuppam Beach. There is a gentle nudge now and then from the southwest. The clear morning ocean leaves long pauses between the waves. The sky is vacant except for a thin crescent moon and a bright Venus near it. At sunrise a Jerdon's bushlark begins its mellow whistling sitting among the ipomoea creepers. Then the soft trills of green bee-eaters perched on the bare thorn branches of wind-beaten prosopis. The air is so still that a group of crows dares to wander more than 50 metres into the ocean, flying around friskily, not with any seeming purpose but just for the novelty of it. That they can finally fly where only the deft of wings like the terns and gulls can.

Most boats from Urur Kuppam had gone into the ocean around 2 a.m. to cast nets for *kavala* meen (goldstripe sardinella). Some returned with a decent catch, others empty-netted. The ocean is going through a dry spell. Gopi, a friend and fisherman, describes to me a sardine school at night as an 'array of mirrors rising up and shimmering, making your eyes tickle'.

Like so many other fish, sardine is more abundant at night. Their movements follow that of zooplankton—a vast assembly of microscopic animals, from the larvae of crabs, barnacles and sea snails to all kinds of organisms of the strangest forms when seen magnified, hole-riddling our ideas of what life can look like. Such as the radiolarians, actinarians and foraminiferans. They rise to the ocean's surface at night to feed on phytoplankton; descend into the twilight zone or deeper at sunrise to elude predators. But fish and other creatures have learnt to follow them. Plankton churn the massive ocean columns across the world every day.

It is the largest animal migration on earth and an unseen gear of the biosphere. Diel vertical migration,[2] their daily movement is called. To me this feels like the ocean dreaming. An upwelling from deep in its being every night, playing out mysteriously when most other life is quiescent. Like our own dreams, when things from deep inside well up and play out, making it difficult for our wakeful self to make meaning of them. Most of the mind is subconscious and barely understood. Most of the ocean is subsurface, far beneath its sunlit zone, and barely understood.

When the sun climbs about 15 degrees above the horizon, a steady southwest breeze begins to push. In summer the land–sea breeze inversion is quite stark and by afternoon the wind steers around and blows from the southeast. Two courting carpenter bees fly up together into the sky, gleaming like dark shooting stars, till I can see them no more. I come back home with a two-toned face, feeling the heatwave which is now upon India, the fourth one in the last two months.

At 4 p.m., south of my apartment, I notice the Perungudi dumpyard ablaze. It is part of the Pallikaranai marsh, now proposed to be a Ramsar site, but nearly 200 acres of it is a dumpyard for Chennai. The massive smoke columns are at least five times taller than the pylon towers parading through the marshland. The Chennai corporation says it is a methane fire—usually caused by unsegregated organic waste, also called legacy waste, sparked in hot weather. The fires can go on for many days. The dumpyard in Delhi too is burning for the same reason; of course, there the mercury is almost touching 50 degrees now. Breathing is difficult later in the evening and we shut all our windows. Smoke plumes make clouds over us, shroud whole buildings, create a bright orange sunset. Night herons, ducks and waders fly around in the sky in panic circles all evening, calling loudly, fleeing from the suffocation.

[2] Andrew S. Brierley, 'Diel Vertical Migration', *Current Biology*, Vol. 24, No. 22, November 2014.

Perumbakkam/Sholinganallur Marshland
29 April 2022

Rain tree flowers fall in heaps and are gathered in the unswept corners of the apartment park. Tangled masses of brown, shaggy tufts, as though bears have been shorn.

Indian ash stands in its summer dreadlocks. Its hanging flower racemes are made for the smallest insects around—odour ants, longhorn crazy ants, stingless bees, tachinid flies, parasitoid wasps and so many UFIs (unidentified flying insects). The largest creature which comes to ash flowers is a paper wasp, which perches on the tiny white flowers—whose stamens stick out like sea urchin spines—with difficulty.

At 6 a.m. I leave home to see the flock of flamingos which have been hanging around at Sholinganallur marsh for a week. For my mother it is the first time seeing these birds. We count about eighty of them standing and foraging in seven or eight groups scattered across the wetland. Many young ones seek solitude and forage alone—their bills and feathers yet to turn bright pink. One sub-adult bird flies to about 100 feet from me, pedals on the black slush and sifts it rapidly. A black-winged stilt comes by and follows it around for flushed food. Then a couple of small egrets. They are all comfortable with each other's company.

Land breeze brings the thick smell of marsh bottom being stirred by flamingo feet. The tall adult birds call like ponies. Their sociality bears stark semblances to ours, to the extent we know them. They raise their chicks in crèches cared for by a few guardian adults. They hang around together, mostly in distinct age-classes—adults together, sub-adults in a group, the very young ones usually preferring to be alone—often in the company of other birds. Flamingo pairs are known to form between birds of similar age. Among human adolescents you see a similar pattern, when changes in the temporal lobe of the brain cause what we could call 'non-peer blindness' making young people super-focused on their age group and spontaneously ignore the significantly younger and significantly older. If we find sociality in a different galaxy we'd possibly see patterns that are immediately recognizable.

The light on the birds is a strange orange, scattered by the fire smoke from the dumpyard stretching across the eastern sky. New multistorey buildings are closing in on the wetland. The hills to the south are almost invisible now.

I am a petitioner for the first time at the National Green Tribunal against the two mechanized harbours proposed at the Kaliveli Estuary 100 kilometres from here, for which illegal roadworks and dumping of debris in the marshland began even before the fisheries department got consent from the Pollution Control Board.[3] Waterbirds and migratory shorebirds come in their thousands, and olive ridley sea turtles nest along the beaches. I ask for solidarity from the pelicans, ibises, waders, sperm whales and turtles to protect the wetland and the sea.

Chennai
18 May 2022

Some solitary bees sleep by biting a stick, vine or blade of grass and hanging on it all night like clips on a clothesline. Among them are the green-eyed, fuzzy-bodied blue-banded bees (*Amegilla sp.*). Their backs and legs are furry like Labradors.

Some other freelance, hiveless bees sleep by hanging from a bare twig on their forelimbs or with all limbs, like a child on a monkey bar. Sweat bees sleep like this, in two rows facing each other in a zip-teeth formation.

---

[3] M. Yuvadeeban vs Department of Fisheries & Ors, Appeal No. 14 of 2022, Before the National Green Tribunal, Chennai. (In September 2023 we won the appeal.)

Two days ago I was at Gerry Martin and Chandini Chhabra's farm with my team of naturalists to understand Gerry's work in reptile conservation and Chandini's in nature education. On a tiny dry stem of a duranta plant I saw about forty-five bees asleep at night. At the twig's base were sweat bees crowding together; to the tip were blue-banded bees biting and holding themselves in a slanted cantilever; in some places both species slept mixed. Some sweaties just plopped themselves over their sleeping kin, without a need to grip anything. Flashes, flashlights, murmurs made their antennae twitch, heads and wings shift. Darkness and silence let them sleep again.

A scops owl glared airily back at us nightwalkers from a low mahua branch, with a raptor's boldness and confidence in the night.

For the last few weeks I've been keeping a dream journal. A small notebook which I keep by my side when I sleep and summarize the dreams I can remember when I wake up. It is a little sonar I am trying to use to gauge the deep sea of the mind. To observe without any interpretation. An effort to 'make one's darkness conscious', as Jung put it, or just to pay attention to this extraordinarily ambivalent and non-linear field that is the dreamscape. Jill Mellick, a Jungian psychologist, wrote, 'Hold your dream as you would hold a butterfly—in your open, quiet palms.'[4] I am also listening to dream scientist Gina Poe who has studied the varying nature of dreams during different stages of sleep, when different parts of the mind are in conversation with each other.[5]

I began the practice after noticing that a certain nightmare stopped recurring after a set of events, leading to a phase of

[4] Jill Mellick, *The Art of Dreaming: A Creative Toolbox for Dreamwork* (Conari Press, 2001).

[5] Dr Gina Poe, 'Use Sleep to Enhance Learning, Memory and Emotional State', Episode 30 of the Huberman Lab Podcast.

emotional growth. I've learnt through dream journalling that when I tell myself to remember dreams, I remember them.

On the first night of seeing these sleeping bees I had a difficult dream—of taking pictures of an environmental offence inside a power plant in Chennai and then running from attackers to protect the evidence and myself. I went to hide in a classroom at my old school, called up my mentors, teachers, family for help. A recurring pattern in my dreams. Recording it made me think about what activist friends experience in their dreams, and how campaigning constantly may shape our deep psyches.

I asked my climate activist friends Disha Ravi and Sriranjini Raman about their dream realms and what they face there due to their work. Disha told me that after an intense day of doing climate campaigning, she had intense dreams. She recounted a specific dream where she and a friend did a covert operation in a mining facility and made a documentary, but that did not prove powerful enough to shut the operation down or move the government. She described feeling overwhelmed and powerless, but there was also collaboration and solidarity with friends. She sometimes experiences shapeshifting into the creatures she is trying to protect. Sriranjini shared repeated instances of dreams wherein friends were getting arrested for standing up for wild places or for writing, singing, making art or sometimes dying due to the climate crisis. Other friends shared troublingly similar stories.

Trying to understand this more I discovered the work of Martha Crawford, a therapist who collects dreams about climate change,[6] and Sally Gillespie, a depth psychologist who explores the impact of the climate crisis on our collective subconscious.[7] The latter

[6] See climatedreams.com for the Climate Dreams Project.
[7] Sally Gillespie, 'Climate Change and Psyche: Conversations With and Through Dreams', *International Journal of Multiple Research Approaches*, December 2014.

says that 'dreams in particular served to highlight paradoxical and confused understandings and responses which arose through our embrace of dualistic conceptions such as life and death, neglect and care, hope and despair, acceptance and denial, individual and social', while psychologically coping with the climate crisis.

Through watching my own dreams I am learning to be comfortable with ambivalence, and also that the deep mind does not see things in polar oppositions, it moves easily into paradox, which the wakeful mind finds perplexing. Dream-watching also shows that landscape change inextricably creates mindscape change. I see this from my own practice as well as from listening to friends.

The next morning is rainy and the sweat bees at my balcony decide to sleep in for a bit longer, almost till 9 a.m. On sunny days they are up by 6.30 a.m. or 7 a.m. I think about what bees may dream. Do their dreams change with the changing landscape— early monsoons, lesser flowers and stronger storms? Do their nightmares have smoke, spider webs, brush cutters and spray cans? Is our collective unconscious coupled in some way to that of all other creatures living at this time?

For the past few days a magpie robin has begun to sing by 5.30 a.m., its song so clear in the chill air. It seeps into my dreamscape for several minutes, bringing lightness and reassurance to me in that realm.

Velachery Marshland
19 May 2022

The first sunny morning after several overcast days. Clouds had pushed the black kites down till they could glide only along the neem treetops and be perpetually bullied by crows. Their wings aren't made to flap for too long this low in the troposphere.

The sweat bees in my balcony pots have been extremely busy, flying in and out, in and out, always returning with pale-yellow pollen on their hind legs. To watch them I have to sit hazardously close to the paper wasps nesting in the vacuum cleaner bag. When I do, a few wasps climb out and perch on the cover to keep an eye on me.

The fallow land in front of the apartment has been cleared for a driving school. It has been days since I saw a mongoose. No longer do I see it spring across the grass in the mornings and evenings, making waterhen, squirrel and francolin scold and swear.

Yesterday afternoon I went out to track down the flowers the bees were going to. I found the copperpod tree on the street cloudy with sweat bees. Every inflorescence had ten to fifteen bees. The whole tree might have had more than five hundred of them harvesting pollen, flying in from all the garden pots in the apartments nearby.

Later in the afternoon I stand outside a mobile showroom waiting for my mother, watching about fifteen little swifts in the sky. They sail 150 feet above the hectic road, the half-done bridge and the bus stand. They are 'watching over' us in a sense. The birds pour upward into three empty balconies on the ninth floor of a blue-glass apartment. They have their nests there most likely, made of mud and pigeon feathers. They come out to fly like eddies and swirls in a stream, bifurcating, trifurcating, scattering, converging, flowing in the sky under a coral reef of clouds being sucked seaward. Even in the din and cramp of traffic, there is ever always the sky to look at and the possibility of swifts. My mother says that's why she won't let me drive the car.

The golden showers tree is in full stormy bloom near the train station. Ears want to hear a gushing, inundant sound as eyes see their hanging flower fountains. Yet aren't they flowering too late in May? Three days ago the news reported that the southwest monsoon might be early in the Western

Ghats. The mass flight of termites augured it when I was there in those hills recently.

Opposite my study table in the hall is my sister's desk. Under it we've kept her large black makeup box next to her schoolbag, still with books, which we don't have the heart to unpack even years after her passing. A potter wasp (which we've named *kuyava kulavi* in Tamil) is bringing clay from outside and building her nest on my sister's bag. She has kneaded together three little pots the size of a baby's thumb on the thick, furrowed plastic surface, with a neat opening that has a lip on top. A little while later she flies in carrying one by one green caterpillars of geometrid moths, sticks them into her nest-pots and seals them.

Difference Meditation
Kovalam / Muttukadu Backwaters

Yesterday I accompanied my students from our Youth Climate Internship into the grasslands near the Kovalam backwaters. We were on a snake walk with the Irular tribe. When you meet a snake, sometimes the first thing you feel is 'difference' and the creature's sheer otherness. How it sees the world, how it leads its life—your human body squirms inside if it tries to imagine what it is like to be a snake. I saw these reflected in the questions the girls and boys asked Kali Anna, who led the walk. Do male and female snakes have different temperaments? Do female snakes have periods? Will a venomous snake slowly become friends with me if I keep it and take care of it? Each doubt—seeking to discover similarities with this wild creature—deepens our minds to its differences. About 200 painted storks and pelicans rested lazily on a sandbar in the backwater for all the five hours we walked in the grasslands. It became a really hot day once the dawn clouds cleared, but there was skylark song pouring from above throughout. I came home, slept, then got up and wrote a meditation on 'entering difference' to help me consciously acknowledge and step into the otherness of life around me.

(Entering difference is a guided meditation to be read slowly, with sufficient pauses between lines, and longer pauses between paragraphs. To be done in the shade or indoors, preferably sitting. You may replace some of the features in it and bring in creatures living around you, wherever you are.)

Gently close your eyes and tune into your senses. Tune into what lets you perceive the world outside your body. The entire surface of your skin which lets you perceive what is immediately outside your body and touching it. Then your sense of smell and of hearing which let you perceive the world at further distances. Tune into all of them together and stay aware within them for a minute.

Now bring your attention to the senses which let you perceive the world within your body. From head to toe, feel the volume of your body and what it brings to your awareness now. Bring to mind the joys and aspirations you presently have. What purpose fuels you at this time in your life? What gives you peace? Bring to mind any fears and worries you may have at this time. Observe them keenly.

Now imagine slowly all of these feelings withering away from you. Imagine the vanishing of your senses and memories. Imagine your body shedding these and transforming into something entirely new. Imagine it transforming into a shorebird—growing wings, feathers, a beak and talons, like a sea eagle or a tern. Feel your body now. Imagine how you will see and feel the world—the winds, the ocean and the sky. What might you perceive now that you couldn't in human form? What can you no longer see? Become aware of the world inside your bird-body—what can you imagine feeling, as you perch, take flight, soar or sit in your nest? Now think about what joys and urges you might feel as a shorebird. What aspirations and purpose fuel you?

What fears and worries might you have? Observe them keenly for a minute.

When you are ready, shed this body too and imagine plunging into the deep sea where no light reaches. Imagine yourself now as a deep-sea creature, like a lantern fish or an angler fish. Imagine living in waters heavy and eternally dark and cold. The sun is like a distant star far off in the sky. Grow and feel your gills, fins and scales. Imagine how you'll see the world. What will you pay attention to? Enter your body and imagine what you might feel inside, what this new form might let you experience. What will give you peace and purpose? What will be your urges? What fears will you have? Observe these for a minute.

Now imagine yourself gently rising onto land and changing into a human again. Not into yourself, but somebody you know, whom you meet often but is extremely different from you. Imagine being in the mind and body of this person who is entirely different from you— different social background, different lifestyle and beliefs, different aspirations, culture and so on. Imagine being in this person's body as honestly as possible, shedding your own thoughts and beliefs. What will the world seem like, what will you pay attention to? What will give you peace? What will be your fears and worries? Try to observe these as keenly as possible.

Now transform for the last time and settle into your own body. Feel its uniqueness. As you settle into your body, become conscious that your perception of the world is one among billions of ways there are to see and live on earth—each as unique as the other. As you settle into your body, grow aware of its energy. Become conscious that your body and mind are a community themselves of billions of different living entities working together. Become conscious—through feeling your living body—that difference is a necessity for all life and your life. And that the earth holds together an uncountable number of ways to live, to perceive and to be—each so utterly different from the other.

Sadurangapattinam

22 May 2022

A male koel's is the first call at 4.47 a.m. A sudden splintering of the dawn's silence as though the night has been quiet for unreasonably long. The fruiting ash tree below the balcony is its outpost till the crows chase it off vehemently for waking up everybody early.

Sadurangapattinam (Anglicized as Sadras) is a coastal village 70 kilometres south of Chennai where a small distributary of the Palar River splits from the main stream and meets the sea. When you stand on the beach, to the north you see the massive vapour stacks of the Madras Atomic Power Station, and to the northwest stand the ruins of a Dutch fort built in the thirteenth century.

Seven of us naturalists started in the morning, came to and walked along this stretch of coast as the fishing boats returned with heaps of crab nets twitching with tongue soles, three-spotted swimmer crabs and Bengal whiprays. The nearshore waters were a clear, greenish blue and didn't hold much fish. Offshore fishing is banned till 15 June for the fish-breeding period. A steady southwest wind stroked the curling wave crests, plucking out froth in long silk strands that vanished in the hazy air.

Low tide is at 6.56 a.m. Millions of small shells and their fragments form strandlines, outlining the edges of waves when the tide is higher, marking their surges and crashing shapes. I wish I could read into them more finely, and be able to tell something about the ocean and ocean bed from their curvy script and constituents.

We need a phrase in our ecological and political speech like 'nuclear mistakes'—*karu thavarugal* to neologize in Tamil—those which once committed cannot be undone in planetary time. Walking here I think about the subsurface realities of people living near a nuclear plant. The Madras Atomic Power Station in Kalpakkam is riddled with a history of leaks—in 1986, a blast of deuterium from one of its units, then in 1999 another leak exposing forty-two workers to large amounts of radiation, then in 2003 again dosing six workers. Accidents are called 'incidents'[8] in nuclear science language, till they are explosions.[9] Like deaths are called casualties. Casual for whom?

Indian law hushes up proceedings in nuclear stations from public reach by calling them 'strategic' and exempting them from various regulatory safeguards. A journalist friend has been investigating the prevalence of several illnesses in these surrounding fisher hamlets near the plant, but procuring and publishing that information is proving impossible. Since 2021, land within 5 kilometres of this plant is being sold, within its 'sterile zone', both against law and good sense, putting down the foundations for a future disaster. Radioactive nuclei are deep-time entities. Uranium's half-life is 700 million years. This coast will remember some mistakes for all time.

In the mouth of the beach-blocked stream, river jellies (*Acromitus flagellatus*) drift with thick, stubby tentacles. Their bells have fine red dots and some are digesting anchovies that can be seen through their transparent bodies. Fishermen tell us they glow phosphorescent green at night.

When the southwest monsoon begins on the west coast, branches of its winds travel to Tamil Nadu. How can we know? Dragonflies. Small swarms of wandering gliders, along with picture wings, are gliding over the beach vegetation. In May and June they catch the monsoon winds and travel along the west coast into Central Asia. But some come here this season, and I am excited to see them after three months.

[8] T.S. Subramanian, 'The Kalpakkam "Incident"', *Frontline*, August 2003.
[9] S. Anand, 'India's Worst Radiation Accident', *Outlook*, February 2022.

We set off to find the Palar River's estuary a few kilometres south. But we are wrong in the way we imagined the river. The Palar reaches the sea like a jellyfish with numerous tentacle streams braiding around and into several large sand islands the river has made on its lower course, which have become scrub, forest, casuarina plantations and farmland. Around this giant dreadlocked river mouth the 2004 tsunami still starkly marks many villages. Along the intertidal are battered and broken fish auction halls, strips of sea-eaten houses, and fisheries offices and storage rooms that are caving in.

Sadurangapattinam is also famous for the old Dutch fort in it. Inside its cannon-bombed, battle-worn walls, I see nature quietly reclaiming space, undoing what was war-made. The ammunition rooms now house mud nests of little swifts on their ceilings and burrows of digger wasps and antlion larvae all across their sand floors. Three spotted owlets roost on a neem tree, cracking through a standalone stairway starting nowhere, leading nowhere, hanging mid-air and turning into tree. Bee-flies—the most graceful of insects—nectar on oldenlandia flowers covering the graves in a cemetery of Dutch dignitaries, along with moss and grass.

Theosophical Society
30 May 2022

I visited the Great Adyar Banyan again last week. My nature-educator colleagues and I had taken children from Urur Kuppam and Ramapuram on a walk inside the Theosophical Society—learning how to read clues and tracks on the sand, the trees and through the soundscape. The children loved playing with the utter hydrophobicity of lily pads in the pond near the tree. We listened to the afternoon's calls—the flameback woodpecker, treepie, squirrel's alarm at every passerby—as the day touched 40 degrees. Not next to the Great Banyan though, this being which had been alive for over half a millennium.

When the nosy Portuguese sailor Vasco da Gama reached India, this tree was a seedling girdling a palmyra or a neem. Five centuries later, its prop-root web is now wide as a village, its phyllosphere large as a monsoon cloud. During a storm in 1989 its main trunk collapsed. But the tree lives on, anchored by its aerial roots, each like a small tree trunk. A banyan is known to 'walk' around with them, its strides stretched over decades, dropping roots towards where it wants to move and withdrawing them in other places. This one is possibly related to, mother of, great-great-grandparent of almost every other banyan in Chennai. In front of this tree is a trilithon—a stone gate known to be over 2,000 years old—placed here like a portal by Colonel Olcott, one of the Theosophical Society's founders.

The last time I visited this tree was three years ago, before COVID-19. I had come to it for consolation, three days after my sister died, and spent some time under it. I could cry to it safely, a grief I had not poured anywhere. Banyans enter my dreams easily, without asking. Their roots and folds and chaotic connections are too nerve-like, thought-like, dreamlike. They reach into the mind which sees itself as it sees into canopy.

That night I walked on my street, my home and other apartments gone, banyans in their place. Grass overgrown, a playground sinking into wood, hawk-eagles overhead with wingspans too large. Then a mongoose crossed my dream path, from right to left, saw me face to face, then trotted along.

Friendships formed across species allowed leaps in life's evolution, showed Lynn Margulis, among my favourite evolutionary biologists. Friendship between eukaryotic cells and bacterial cells created the animal cell with mitochondria and the plant cell with chloroplasts. Friendship between plants and fungi entirely changed the atmosphere's composition. Between polyps and zooxanthellae in the seas making coral reefs, then between flowering plants and bees, and so on. Symbiogenesis, she called it.[10] I saw another symbiogenesis remaking the world right around me, silently transfiguring the human-made world, brick by brick, every day.

In the morning I sat on the balcony and watched for a while the peepal sapling growing from the railway station's pillar and the 3-foot banyan on its sunshade. Later I walked into the unoccupied spaces of the station towers and noticed ficus roots cracking the withering cement under the station's terrace, running alongside the rusting steel bars.

Some ficuses—like banyan, peepal and bat fig—will never germinate on the ground. Koel, oriole, mynah and other birds must eat their fruits and leave their seedy droppings on another tree. Then that ficus grows like an arboreal octopus into the host tree's trunk, sucking out their health, growing down, rooting into ground. Buildings too hold nutrients—brick, lime and sand— which the ficus has learnt to tap. A wet, mossy patch on a roof or sunshade is ideal. Or just sheer vertical wall. Steve Cutts in his film *MAN* shows a dark, post-human landscape of cement and skyscraper wastelands, lifeless and uninhabitable, upon which a lone white man sets his throne. It's a powerful image but it may be wrong. A post-human world may well be a dense ficus forest forged from brick and concrete, planted by frugivorous birds, full of life, people included, perhaps.

On the weekend Rohith and I go around old Madras to see the slow insurgence of ficus trees. Given fifty years in their space, they will climb over our monuments and dismantle them. They see no need for old British forts in the city. Might as well turn them into trees. Ficus roots are breaking open prison chambers and the armouries at Fort St. George, cracking the corners and climbing out of its war moats. Old bungalows of merchants and zamindars in Washermanpet are already four-storey forests. Peepal pours

[10] Michael W. Gray, 'Lynn Margulis and the Endosymbiont Hypothesis: 50 Years Later', *Molecular Biology of the Cell*, May 2017.

out, roaring out of French windows, its roots run like and into plumbing. The Central Press' old colonial buildings are also run over. Invading banyans have been chopped off multiple times but they've grown through the floors like veins, such that a new building has been built for the press to function.

Parakeets and mynahs fly out from the branches inside Indo-Saracenic balconies. Tailorbird and ashy prinia call from inside a warehouse. The apex towers of the high court host ficus seedlings. One tree has knocked down the wall of an Amman temple built around it, and the place is enclosed in green sheets for renovation. Abandoned agraharams north of NSC Bose Road have huge bat fig and banyan breaking out of their tiled roofs.

Temples, churches, mosques—any religion can be turned into bark, leaves, heartwood and sapwood.

Nochikuppam Beach
2 June 2022

At 6.30 a.m. the air is warm and the sun is sharp. The gathering land–sea breeze combs the curls of waves into long, static strands seeping into vapour straight up, as though a Van de Graaff generator is at work. A sight only seen in summer. My eyes don't let me move for a while. A lone palm swift, far from any tree, is inscribing some loopy language over ocean air with its perfect pickaxe body.

Plumes of rising froth may carry with them micro-constellations of aeroplankton—plankton lifted into the breeze and floating there, I imagine, in their strange shapes. Dragonflies are known to hawk them. Possibly swifts do too.

My friends and I walk to Nochikuppam, south of Marina Beach, to document fish catch this summer. Fisherfolk call this the marappu time, when the ocean is blue and clear and bad for fishing in open water. Many fishermen go to the reefy and rocky areas, an underwater terrain they map not through sonar and scuba-diving but through intergenerational observation of the species composition in different places. Fish species speak a lot for the ocean bed, its habitats, currents and health. Indian mackerel in large shoals, for instance, forerun the northeast monsoon current. Slipper lobsters and other large crustaceans caught nearshore may signify upwelling events, what fishers call olni. There are several other patterns of interspecies associations; for instance, good prawn catch is supposedly indicative that it's also a good time for squid.

I drove through the market yesterday and spotted on one of the tarpaulin-sheet stalls a blue-ringed angelfish (*chippili* in Tamil), its stripes still flickering. Today we saw a whole bunch of reef-dwellers in the market—vagabond butterfly fish (*ookiri*), blue-lined rock cod (*vari kalavan*), rabbitfish (ora meen), blue-striped snapper (kal sankara), yellowfin surgeon (koli ora), among several others. There are places in this ocean which have natural underwater hillocks, such as the Thiruvanmiyur kallu. There are stretches where successful coral reef restoration has been done by the M.S. Swaminathan Research Foundation, where good numbers of fish congregate. After chatting with Palayam Anna about reef fishes I learnt about another small nearshore reef which is popular with fishers.

Vandi parai (vehicle rock in Tamil) lies about 40 feet underwater on the ocean bed a few kilometres from the beach of Neelankarai/Kottivakkam, and according to him there is a blue-ringed angelfish that can only be seen there. This reef area came into being when in the 1960s a Hawker Sea Hawk fighter jet from the Indian Navy crashed into the sea.[11] A young crew member took it for a joyride and didn't know how to operate the plane. The pilot was saved by a fisherman, but the jet sank into

[11] Kamini Mathai, '55 years after Grandad Saved Pilot, Chennai Diver Locates Jet Wreck', *Times of India*, February 2019.

the sea. It is a small craft—about 18 feet long, but somehow it has aggregated into a reef which is a hotspot for several species.

At Nochikuppam, beauty is an oppressor. Beauty is classist and brutal.[12] This road by the beach north of the Adyar Estuary was laid in the 1970s, for upper-class people like judges, ministers, administrators and others to commute in their vehicles to Fort St. George, the high court and other government buildings. Fisherwomen from around twelve hamlets south of Marina Beach sell their fish by the roadside, on what was once their home beach. Time and again through the years different parties have been vying to acquire this stretch—not a habitat in their eyes, not comprising livelihoods, but prime seaside real estate.

In 2019–20, a judge who had to pass through the loop road found the fish markets an eyesore. He wanted a road joining Besant Nagar and Pattinapakkam leading into Nochikuppam, so that his travel time to the high court could be reduced.[13] This would also require evicting the fisher hamlets on either side of the estuary, which in his eyes would 'beautify' the place. Such a road would also endanger the turtles and other life during the migration and nesting season. We folks at the Chennai Climate Action Group joined hands with the fishers and campaigned till the ₹400 crore project got shelved. When a city is not made for human beings but for the movement of private vehicles (and those in them), its idea of beauty becomes ghastly and oppressive. Presently a mall and business centre, along with rope cars, jogging trails and sidewalks, are proposed here, and the word 'beautification' is overused in the documents. In none of these plans are the fishers featured, let alone consulted.

[12] Shobana Radhakrishnan, 'Why Fisherfolk in Chennai Are Opposed to Beach Beautification Projects', *Citizen Matters*, October 2022.

[13] *The Hindu*, 'Loop Road–Besant Nagar Link will Cost ₹411 Crore, High Court Told', 4 December 2020.

Stories like this abound—tacitly revealing who matters and who doesn't, what is beautiful and what is not, what is dirty and what is clean. At the Chennai Climate Action Group we work to termite and tube-worm these ideas in the public's mind. The invasive prosopis bush was brought into India with two main agendas—to provide fuelwood and to 'beautify' the Thar Desert. The plant has now destroyed vast swathes of native habitats, displaced pastoralists and is learning to thrive on the seashore too—all of them 'beauty-impacted' landscapes. In the late 1990s coal being handled at the Chennai port threw up soot which reached a high court judge's windshield on his way to work, making it dirty. He got irritated and ordered that a new port be made for coal-handling. Hence, the Kamarajar Port was established on Kattupalli Island, evicting, obstructing, affecting dozens of fisher and coastal villages and decimating the ecology of the sand-barrier island, which is among Chennai's most important climate guardians.

Kallu Kuttai
12 June 2022

5.15 a.m. The sky is a planet-sized pair of 3D glasses. The northern square is pink and the southern square is bluish-violet. Venus is at about 40 degrees in the southeast above the horizon of IT buildings. Schools of young tilapia fish (*jilebi* meen in Tamil) create reverse ripples at the Kallu Kuttai Lake, like handfuls of sand thrown over the water.

I have been visiting this shrinking wetland, a few minutes on foot from my home, since 2016. It has been cleaved by a road linking the Perungudi and Taramani railway stations. But Kallu Kuttai swallows the road between October and December.

Every morning you can see about 300 pelicans foraging tilapia. But not later than 9 a.m., by when they would have flown off in batches towards the Pallikaranai marsh to the south. For some years I had imagined that Kallu Kuttai was a midnight roost for these birds. Since you cannot spot a single pelican even at 7 p.m., I thought they fly in after dusk and spend the night. However, when I went to the lake at 5 a.m. this morning, I saw them gliding in, in small flocks from the west. Only one flock of fifteen came in from the north. There were already about ninety birds when I arrived. By 6 a.m. there were over 200 foraging with half-lifted wings and plunging beaks in two long, disorderly lines. The morning sun drew a golden median across the water perpendicular to the feeding pelicans. I learnt that these birds visit this site in their daily travel circuit, a stopover early in the morning, for about three or four hours.

When pelicans glide in the dawn sky, the wedges they form are like flowing ice, only slightly fluid. Their wing beats move from bird to bird in a slow pulse. The flock retains its shape impeccably even as the birds sail above. Watching them has the same effect on me as many minutes of meditation.

Dawn break ...
ripples from a pelican
touch the shore.

Two pelicans hunting near the water's edge where I sit paddle further away from me. Together they create a widening W on the water's surface which ripples onto each shore. I see for the first time courting cloudwing dragonflies, spinning up into the air in superfast double helixes. Six gigantic cranes look over the lake from the northeast, each holding its metal neck in different postures. They are all perched over the new multistorey DLF building under construction.

Several sewage outfalls from the big apartments go into Kallu Kuttai. That is the bittersweet beauty of Chennai's wetlands. They host astounding life even after decades of abuse. By 6.15 a.m. my bird list is as follows:

S.B. pelican: 200+
Cormorant (lesser/Indian): 30+
B.W. stilt: 6
Darter: 8
Dabchick: 6
Pond heron: 11
R.W. lapwing: 2
Little egret: 24
Night heron: 2
Pied kingfisher: 2 (passing overhead)
Glossy ibis: 60+ (passing overhead)
Plain prinia: 4
G.H. swamphen: 2
Eurasian moorhen: 2
Purple herons: 3
Painted storks: 5
W.B. waterhen: 2 (call)
Yellow bittern: 1 (call)
W.T. kingfisher: 3

Later in the morning under the railway bridge, women from Kallu Kuttai Kuppam are selling catla, rohu and tilapia caught in the lake. Sometimes I see snakehead, Eurasian carp and lake pomfret too.

New tenants are coming into a nearby apartment and are shifting their belongings in large brown cartons. About fifteen crows have gathered curiously around the packers-and-movers vans. They are nesting on several electric cables in my neighbourhood. In all the nests, they have sewn together sticks with white packing tape.

Urur Kuppam Beach
15 June 2022

Sitting under the shade of a boat, I watch the waves. A pale-brown stray dog with bite wounds on its ears and mange on its back shares the space with me, curled half asleep in its sand crater. Waves spilling and sliding down on themselves hiss, like water splashing on a hot pan. Waves which plunge and curl over, the slightly taller ones, roar as they rise and crash thunderously as they fall and bounce. There is a near-constant mix of crashing, roaring, hissing; hissing, roaring, crashing. Occasionally there's silence, at most three seconds long. Tower snails roll back into the falling tide with each receding wave. They live almost in perpetual rotation up and down the subtidal slope.

A crow examines the ear of a pink teddy bear lying with its face to the sand. I think about how other creatures navigate and investigate the unknown and the unknowable. How do they respond when they come upon something which is utterly without analogue in their minds? Perhaps they gawk and get curious too. Is perception only about fitting bits of reality into the mind's existing frames? Sometimes not, I think. Having walked the coast enough times, I've seen how a place creates new frames in my mind, wired neurons in patterns which accommodate it. Then suddenly one day something so long invisible becomes a stark reality. Listening to wave sounds, for instance. A crow, then, having spent generations on this beach, must be proficient in the properties of a hundred forms of litter. The tensile strength of a single-use spoon, the usefulness of nylon strips, the risky meal offerings of a ghost net.

The beach's berm has a step. It is more than 2 feet high and about 20 metres beyond the intertidal, carved by the large, surging waves of the cyclone in May which blew in the water up to the ipomoea creepers. The sky today has so many different clouds—wisps, puffballs, sheets—but mostly it's clear. Yet there are some mildly surging waves 4–5 metres beyond the high-tide line, leaving behind pools of water in a few places. Some wave crests follow closely behind one another. The weather is scorching, but the swiping water is cold. Perhaps signs of an upwelling ocean.

Sardine, anchovy and mackerel catch are slightly on the rise compared to the sparsity of the past couple of weeks. One prawn net caught two lionfish (*thumbi*, homonymous in Tamil with dragonfly), a creature which has a venomous spine over its back with cobra-level neurotoxicity. So I request Sudhakar, a middle-aged fisherman, to teach us how to hold it correctly.

I see a zebra blue butterfly flying into land low above the water—with its tiny wingspan of 2.5 centimetres—straight from the open ocean. Some tawny costers do the same but from the southeast. After 9 a.m. the searing coastline becomes a flyway for migrating lime swallowtails.

In August 2017 I had gone into the ocean with other birdwatchers to survey pelagic birds—those species which never come to the shore except to breed, and spend almost all their time out over sea. We saw six species of true pelagic birds—Wilson's storm petrel, Swinhoe's storm petrel, Arctic skua, long-tailed skua, flesh-footed shearwater and bridled tern. But we saw more butterflies than birds tens of kilometres into the Bay of Bengal. This was our butterfly checklist, which my friend Subramaniam Sankar has methodically preserved—lime butterfly, skipper sp., common emigrant, lemon pansy, blue pansy, tawny coster, common grass yellow and crimson rose. Were they all just windblown? Hundreds of them? Rather I think the ocean is a busy migratory flyway also for insects, where the wind is steady and no obstructions sprout from its surface. They perhaps let themselves be carried till coast reaches them.

In March of 1996, the Indian research vessel ORV *Sagar Kanya* was on its course in the north Arabian Sea.[14] When it was about

---

[14] S.C. Pathak et al., 'Insect Drift Over the Northern Arabian Sea in Early Summer', *Journal of Biosciences*, June 1999.

550 kilometres from the Goan coast, moving north, they did an insect-trapping exercise far inside the sea—those migrating, wind-carried and there for reasons we don't know. They found over the ocean 173 species of insects from 47 families.

Kelambakkam Backwaters
21 June 2022

Wading into the grasslands along the Kovalam salt lake, we are about ten naturalists and some teachers and children from Urur Kuppam in search of snakes. Kali, the renowned Irular snake-tracker, and Alamelu, his wife and fellow snake-catcher, lead us into this soft and sandy landscape. The brackish lake beyond stretches over 25 kilometres parallel to the shoreline, fed by the Kovalam Creek.

It has been discovered that when many Indian snakes are relocated and released away from their home territories, they often die. Is the feeling of 'a sense of home' far more solidified in the perceptual world of a snake? A creature for whom most of the time most of its body is 'earthed', such that possibly its spatiality is unsurvivably disoriented when removed from its belly-felt home. Isn't that a kind of embodied grieving? 'Belonging' for it maybe is significantly formed from a tactile map it builds based on what its belly touches. Most of the 15,000 muscles in a snake (there are about 600 in humans) run on its ventral body tethered to its 400 ribs feeling the ground. A striped keelback shrinks into a squiggle by the sandbank of an eri as our feet approach. I see it later pour across grass and into a thicket covered in prosopis thorns, unstabbed, slick over sharpness like muscle-mercury. Later I crouch on knees and elbows to photograph a saw-scaled viper Kali had found in a burrow. In the monsoon saw-scales climb up and cuddle inside the water-grass clumps, and we search for such snake hammocks. In summer they like to find underground spaces during the day. The many-textured ground presses hard on my prostrate limbs; stones, sand, grass and sticks leave tracks on my skin.

On these walks I try to track Kali's eyes as he tracks snakes. A year ago on another walk he had watched a squirrel on an eucalyptus tree and said it had seen a rat snake. The way it gazed, squealed and flicked its tail gave it away. He searched the bushes around a neem nearby and found a large yellow one.

On a small sliver of flattened, smoothened sand on the slope of the Nemmeli Lake, amongst slipper tracks, marks of rain splatter and goat hooves, Alamelu points out to us where a curved belly has made contact. She says it could be a Russell's viper on the prowl at night. Twenty paces away we find the chain-track skin it has shed—a black-and-white copy of the original. Along the bunds of fallow paddy fields—folds of land—we find a common sand boa, red sand boa (my second time seeing one) and a krait's skin. On leaves of morning glories by the lake we see the entire life histories of golden tortoise-shell beetles. They are miniature luxury tankers. Adorned as if for an occasion. Like a leaf-roaming shard of a star. The beetle's elytra (wing cases) are transparent, through which you see its golden-red wings. When it flies its wing cases open out like a hemispherical glass windshield in front of it.

A beetle made nervous will clamp its elytra down onto the leaf, making its body secure from the outside world—airtight, predator-proof, like a capsule over it. No ant can catch its legs and flip it over. It has also got a transparent helmet over its head into which it will retract its gold antennae. An arthropod in a space suit. Harassed any further, the creature retracts its legs, slides off and falls to the ground, its back facing down. Its elytra are invincible in its cosm. The golden tortoise-shell's egg sac is red when freshly laid and turns brown with a button in the middle. Itchy to touch, their larvae are stout little worms of thick, black misshapen hair under leaves.

It stays with me that snakes reside almost always as guests, in spaces built by other beings. Crabs, rats, termites, ants, woodpeckers, parakeets and humans. Ever nomads, thieves, spirits, leaving behind only slight or withering traces. They occupy or share existing spaces, and never really change the landscapes they live in too much.

Far out in the backwaters the bright pink of about sixty foraging flamingos snake in the heat. Three larks are singing simultaneously, and it feels ecstatic. The oriental skylark is a shivering dot in the sky where westerly clouds are passing. The ashy sparrow lark's long-drawn, soft whistling is accompanied by mid-air parabolas. And the Jerdon's bushlark parachutes upon fencing posts, uttering fourteen whistles on its way down.

Velachery Marshland
29 June 2022

5.40 a.m. Thirty swifts are chirping and playing in the purple sky above the Kallu Kuttai Lake. This bird gang stays in the towers of the Taramani railway station. They are like a swarm of monsoon dragonflies, all chaos, bird-like Brownian particles. I watch them from the pavement during our walk as trains rock the neem trees standing by the track's slopes—which have turned deciduous over the last few years due to increasing heat. The evergreen trees become brown before summer, stand dry for two months, only now learning to shed leaves and grow new ones. Three swifts form a line and fly one after another like a paper kite's tail. One by one more of the gang joins and loops around. It is the Snake game from the old button phones made more rapid. Sixteen in ribbon formation under the blushing sky, and more are joining. They swim like a sea snake, looped into slip knots. They are living, streaming Möbius strips.

Only seven pelicans are at the lake today. The Chennai corporation is building a flood culvert and compound wall along Kallu Kuttai, which has shrunk the lake's margins further by a few feet.

When a marshland dries its clayey soil surface cracks into characteristic jigsaw polygons. This is its hyporheic zone—its liver or gut—as Erica Gies calls it.[15] This is the thin, mostly biologically created layer which cleans the water and lets it percolate underground. When it dries this surface captures the tracks of other creatures and holds them. I've been seeing every day over the summer the paw prints of dogs, the feet of open-billed storks and the curves of a keelback on a specific patch of marsh near home. Now the southwest monsoon rains have washed the jigsaw away.

Further down the road we see the furtive ruddy crake and slaty-breasted rail venture out of their reed cover, but not beyond its edge, to probe the mud and forage along the newly formed pools.

A stretch of marsh by the railway has been fenced off by the Chennai corporation and the Rotary Club. They've dumped dredged mud here (which appears to be from an inland lake, judging by the molluscan remains in it) and are planting trees, all with good intentions. Marshlands and wetlands are Chennai's identity. Forest is a mere few per cent of its land spread. Blind afforestation here is identity erasure. But to the general public forests are more appealing. They yield better funds and show their utilization. These saplings may not last even two monsoons. As per an Intergovernmental Panel on Climate Change (IPCC) report, tropical forests store about 240 tonnes of carbon per annum. Wetlands store 700 tonnes, almost thrice as much as forests. And the most significant carbon storage happens through undisturbed soil and subsoil processes in any habitat. Vegetation

[15] Erica Gies, *Water Always Wins* (University of Chicago Press, 2022).

is only secondary. How and why have marshes maladaptively morphed into barren landscapes in the public's mind? My practice has taught me that barrenness is always a state of mind, never a state of land.

There have been hundreds of tiny wriggling screws in the tadpole tank for the last couple of weeks. I'd have fallen in love with their shape and movements if I had not known they were mosquito larvae. The pupae are oval buns which can still dart around, propelling themselves with their curled abdomens. There is almost no other life yet except a lone checkered keelback which dives into a submerged heap of concrete when it sees me.

Krishnaveni, our cook, tells me that her grandson had contracted malaria a few days ago. She and her family live in Kallu Kuttai Kuppam, an informal settlement spread over 300 acres between the Kallu Kuttai Lake to the north and the Perungudi Lake to the south. E. Deepa, a researcher from the Madras Institute of Developmental Studies, did her PhD thesis a few years ago on the realities of women from Kallu Kuttai who are engaged in paid domestic work.[16] This settlement on the marshland has over 27,000 people and nearly 10,000 families. Krishnaveni, who is in her early fifties, tells me the story of how she came to live where she did. Land was being sold on the marsh, albeit unregistered, at a cheap 3,500 rupees a ground by land brokers while in the rest of Chennai, prices of registered land were several lakhs a ground. This was in the early 2000s when Krishnaveni's son and daughter were in primary school, and she recounts them studying in the light of a kerosene lamp for several

---

[16] E. Deepa, 'Women in the Informal Sector Work: A Case Study of Paid Domestic Workers in Chennai Urban Slum', Department of Women Studies, University of Madras, 2020.

years before electric power was made available. She is originally from near Trichy, belonging to a small agricultural family, and she moved to Chennai when she got married and her husband sought work. Deepa's investigation shows the same, that most of the people in Kallu Kuttai Kuppam came there due to the decline of agriculture and other rural microindustries.

For my third batch of UWW interns I do a field session at Perungudi station, speaking about marshlands, aquatic insects, migratory birds and the settlement at Kallu Kuttai being the human backbone of this entire landscape. Usually when I do a walk there, I show the map of Perungudi, describe the lay of the land and then ask students about the most common species in a marshland: Ducks? Waders? Painted storks? Pelicans? Mosquitoes? No. The most common species in Chennai's marshlands is the IT professional. Along the Old Mahabalipuram Road there are about 480 IT companies, all built on mostly marshlands and lakes. Then arose apartmentscape, which in turn facilitated the movement of rural people into urban settlements like Kallu Kuttai to do unskilled or manual work in these places.

Land there is unlicensed and Krishnaveni tells me about the number of protests they had to mount to get water connectivity, power and other basic amenities. Each time there is talk of flood mitigation, this settlement comes under the radar for eviction. What is not considered during those discussions is that all the IT parks and posh apartments—the main encroachments on marshland—would crumble without this settlement. Yet among the most climate-vulnerable communities in the city is Kallu Kuttai Kuppam, completely flooded during the monsoon with water, disease and neglect. Only recently have some people in this two-decade-old community started building concrete housing, but most live under asbestos sheets. I tell students sometimes that if you want to understand the greenhouse effect, live in a house with asbestos roofing for a day.

In the tadpole tank the first clutches of ornamented pygmy frog tadpoles have hatched and are swimming around. They don't feed on the mosquito larvae but filter-feed on algae and microbes in the water, competing with them. Within a week of tadpole arrival, the mosquito population falls to near zero. There are other predators too now, like the nymphs of granite ghost dragonflies and coromandel dartlets.

I think about the presence of other beings, like backswimmers, granite ghosts and diving beetles—crucial for our own good health in water-governed places, based on what I have observed in this tank over the months. But such a diversity of creatures need naturally clean water. Mosquitoes, I've seen, thrive best in polluted pools and sewage canals, where their predators and competitors are absent.

North Chennai Toxic Tour
1 July 2022

Twelve of us activists from the Chennai Climate Action Group are on our usual route along the coast of North Chennai at the flare of dawn. With us is Sydney Schuler, a young reporter from the Pulitzer Centre who writes stories on how 'development' indiscriminately impacts disadvantaged communities. I sit next to Palayam Anna in the van and show him pictures of fish I took at the Nochikuppam market a few days ago. We argue over fish names. He says that both a swordfish and a sailfish are called *mayil kola* and there is no point calling them different names like in English. I try and fail to understand why.

We cross Kasimedu harbour and the beach vanishes. The groyne fields begin. A number of boats float in the nearshore and men fling *edai valai* (cast nets) into the waves. 'It looks like last night's thunder must have chased sardine schools towards the coast,' Palayam Anna says.

Stop 1: Nalla Thanni Odai Kuppam

Two peregrine falcons flap over the rusty oil silos and fly south along the shoreline. One first and the other a minute later, finding the calm air strenuous to fly in. I see them ascend towards the large cargo terminal at the Chennai port. Vishvaja explains longshore drift and coastal erosion to Sydney. We stand where the village names are mere vestiges—their actual land spreads far inside the ocean due to port-building. Over 800 acres on this coast have been swallowed by sea. Nity points out the strong sulphur whiff of mercaptans the wind brings, and I am astonished by his erudition in identifying gases, their industries and industrial processes of origin and their stories from smell alone.

Stop 2: Burma Nagar Park

Some boys practise football freekicks before a massive dormant smokestack of the Ennore Thermal Power Station—where two new power plants are under construction. Recent protests by people stopped a public hearing a few months ago. The local auto driver is also the football coach. Fifty metres from the smokestack, also under construction, is a thirteen-storey residential complex of the Tamil Nadu Urban Habitat Development Board. Land in this polluted landscape has been arbitrarily changed from 'special hazardous industries' to 'primary residential' for 35,000 working-class/underprivileged people who are to be shifted here from the banks of the Adyar and Cooum rivers. If the power stations ever come up, they will drench these 6,900 families with flue powder and soot every day.

Stop 3: Ennore Creek Bridge

Like dark magic, clouds are made right before our eyes from the coal smoke—flue gas and soot—of the North Chennai Thermal Power Station (NCTPS) and the Vallur Thermal Power Station. Flares burn waste gases from the flare towers of the Chennai Petroleum Corporation. The flame is bright yellow, not blue, indicating that a lot of the toxic gases are escaping into the air. The southwest wind brings a strong stench of leaking LPG. Kotralayar meets the sea to the east and then pushes further north towards Pulicat. Five pelicans glide in through the incessant line of transmission towers and settle on the creek. TANTRANSCO (Tamil Nadu Transmission Corporation Limited) is building an illegal road into the creek,[17] clearing mangroves for another transmission tower. Raju narrates the blockade fisherfolk organized several years ago to stop imported coal reaching the conveyor belts of the power stations, protecting their fishing grounds within the creek.

We look down into the water and take guesses at different fishes, till Palayam Anna tells us they are *sala* and *paranda* (growth stages of *madavai* or mullet). We then hear spine-chilling stories of coal-powered violence from Prashanth—unregulated hot water from the power stations being released into the creek, killing and boiling wading prawn-catchers.

Stop 4: Fly Ash Pond
Raju and Vishvaja tell stories to Sydney of ash-dumping on the salt marshes and the slow death of these rich backwaters. Others check the ash pipes for leakages. Nity finds tide-smoothened pieces of anthracite. I remember reading how some mycologists blame climate change on fungi. If those strange beings had learnt how to digest cellulose a few hundred million years earlier, there would be no fossils or fossil fuels. Trees which fall in swamps today, however, won't make coal.

[17] S.V. Krishna Chaitanya, 'Tantransco Road in Kosasthalaiyar Triggers Flooding, Fishers Flag Flouting of NGT Order', *New Indian Express*, December 2022.

I search for a pair of white-bellied sea eagles which circle these hazy skies and nest through the year on 450 kilovolt-ampere transmission wires. They use the platform they've built to bring food, feed, rest and spend time together, even if they're not breeding. Their eyrie was taken apart last year by a maintenance crew. The birds built it again on a pylon tower a little distance away over the same mudflats. Two large raptors resisting eviction from their ancestral waters. Utter wilderness occupying industry, refusing to leave their home—we have been seeing them over the years raising many generations of eagle chicks over these ash-polluted marshes. The female is slightly larger than the male and can sometimes be seen yelling and ordering the male around. These eagles have a territory of up to 5 kilometres from their nest, made up mostly of wetlands and the sea. They largely eat sea snakes, eel and fish.

Stop 5: Chintamaneeswarar Sand Dune Temple
Behind the temple in the dune's crest is a large peepal tree with several weaver-ant nests which we try counting. Jelly-ear fungi make it feel like the peepal trunk is listening. A dying branch on the tree's extremity has four carpenter bee holes from three of which dark blue compound eyes stare back at us. A coppersmith barbet has dug out the last bee hole for its own nest. From the ocean and over the course of the blocked Karungali River I see dozens of green marsh hawk dragonflies flying in along with marsh trotters. Are they migrating? One marsh hawk sits on the sand with a ground skimmer dragonfly in its mouth, its favourite food, pushing down against spurts of wind that try to topple it over.

I scour through a heap of shells on the backshore collected from the beach sand by shell-pickers for a nearby lime kiln. It is largely bivalves—wedge clams, triangle clams, backwater clams, hammer oysters, green mussels and many species of ark shells. I take one cowl shell, pinkish-red with an umbo, like a sharp human nose. It has a clean, circular hole in it. A sea snail had wrapped itself over the shell in the near past when it was buried in the lagoon's bed, drilled a hole on the nose-like umbo with its acid-squirting radula, and reached in and snapped the ligaments holding the shell closed. Then it had gone in and eaten clean the flesh of the cowl clam. I put it in my pocket and later string some brown yarn through the snail-drilled hole to make myself a bivalve pendant—like the children of Urur Kuppam have taught me.

We breakfast under the shade of palmyras and a swarm of wandering gliders in the dune valley. Three woodswallows chase a soaring shikra out of their airspace. On the way back in the van we tell each other stories of running away from home. There can be as many as four such stories in one bus full of activists.

IIT Madras Campus
6 July 2022

July rains in Chennai have been coming like golden jackals—loud and mostly at night. Like sloth bears, they've been leaving the spires of termite mounds fractured. Inside IIT Madras, under an old tamarind tree, I peer down into the dissolved towers of common snout termites (*Trinervitermes sp.*) with my 10x hand lense. Their large workforce is resealing the mounds with wet mud brought from deeper in their galleries. Soldiers stand around the cracks above the workers with their lance-like orange snouts and their chemical-shooting fontanellar guns. A lost chital deeper in the forest honks loudly in panic, calling to the rest of its herd.

Many ants are termite-traffickers and wait for days like these. Three species of ants are prowling over this mound under renovation. Procession ants (*Leptogenys sp.*) are scouting around in the valleys between termite spires. An ant would stealth in and grab a termite soldier by surprise, especially if it is slightly away from the rest. Termites would stab back at the ant, and occasionally be dropped by its assaulter who then scurries away from the chosen raiding site. Otherwise the ant carries the maimed termite and rushes off downhill and under the thick liana hammocks and thorny *kaattu elandhai* (*Zizyphus xylopyrus*) trunks beyond which I can't follow them. Velvet sugar ants are roaming around but are much more cautious about picking up a termite and getting shot with terpenes. But the tinier and feistier fuzzy pavement ant (*Tetramorium walshi*), about a third the size of a termite soldier, fights without fear and takes away a termite every time. We spot this ant first by seeing curled termites sliding away from the fractures seemingly all on their own, the ant so tiny underneath them.

A few weeks ago I was in Masinagudi in the Nilgiri hills to teach twenty young and accomplished nature educators from all over India as part of YouCan's Earth Educators Fellowship run by my friends Ramnath and Rachita. Before sessions like these I have a practice of walking around in the place and asking the land what I must teach.

Five mountains look over where we stayed. King vultures and rufous-bellied eagles soared above us in the afternoons. Midnight and early-morning rains had dissolved the towers of termitaries. In several places I saw wet earth spewing up ceaseless geysers of termite alates all across the hills. Crevices exhaled gold-winged termites for hours, billions of them. I listed twenty-four species of birds gathering to feed on them. Tree-dwellers like chestnut-bellied nuthatches, yellow napes, long-tailed shrikes and shamas came to the ground and wedged their way between bulbuls, mynahs, magpie robins and spotted doves. A mongoose and a pair of jungle fowl came later for leftovers.

When you stand near a termite emergence, your toes quiver trying to imagine the labyrinths underfoot. On the roads, vehicles struck flying alates, and more birds and squirrels came to pick them off. Procession ant files snaked in the grass, grabbing termites and heaping them over their own burrows. My right leg blocked their path and tingled for hours from the ferocious bites I received from a whole swarm. Wingless termite soldiers and workers congregated above, possibly to misdirect or protect the flying alates from predators. Ants carried them off as well. After the procession ants were done, hunchback ants came to salvage bodies. A marbled balloon frog sat by a bleeding crack in the ground and ate alates for most of the morning, till it could barely drag itself away. Land above was being overwhelmed by life underneath. 'We know so little of the worlds beneath our feet,' Robert Macfarlane writes in *Underland*.[18]

Little swifts and needletails came down from the peaks to fly surprisingly low along with bats, which turned diurnal, and caught those termites which managed to miss a hundred mouths on the ground and rise high enough. A day later I found a mound close to our room clawed out and licked clean by a sloth bear.

Then I saw a termitary, like a mud mound of slow slime, flow up the stairs and onto the boards of a cabin and spread over its side. Wood panels, junk cardboard and window frames in a heap by a storeroom were being silently turned into earth again. And there it felt like the land had responded. Our work as nature educators is perhaps similar to that of termites—to bring down old, rigid structures of learning and reimagine new ways. To eat at and completely hole-riddle the foundations of the existing system. The next day went well. The Tamil word for termite is *karaiyaan*, literally 'that which dissolves other things'.

---

[18] Robert Macfarlane, *Underland* (W.W. Norton, 2019).

Theosophical Society
20 July 2022

I hear my first cicada of the season today, buzzing like a high-voltage wire from a prosopis bush. As it sings, its body temperature is known to rise over 20 degrees above its surroundings. The spit-like spawn of spittlebugs is scattered across the wayside *arugam* grass. Migrating lime butterflies stream in, their flight paths more or less along the roadways, racing vehicles or hitting windscreens, drifting as bright wings and smears of colour on wet tarmac. The bold gold glimmer of picture-wing dragonflies hovering over bushes and tree canopies along with wandering gliders is a sure sign of wetter days ahead. Sea sparkle grows in the ocean near the Adyar Estuary. Crashing waves flare faintly blue, fed by rainwashed river sediment.

I had a banyan dream again, a day after taking a long walk inside the Theosophical Society with Geetha, one of the trustees. A dream scene difficult to describe, aerial roots reaching out all in chaos, my body spinning into their confounding grey and blue braids, stomach tightening and mind losing foothold from lack of comprehension—as if the banyan's anarchist canopy was sedition to all symmetry, rebel to the brain's rules.

During the previous day's walk (in wakefulness) I had seen the front door of an old cottage locked up by a prop root. A young banyan was growing on its roof, devouring its architecture and pouring roots down. One had slithered along the wall, then down the door frame, crept through the latch hole and firmly into the ground. The place was tree-sealed. In another abandoned house by the biggest baobab in the estate—its trunk ten adult hugs wide—a civet lived on the shelves of the kitchen. We entered the house to look for bats in the tiled roofs, ferns in the rain gutters and termite murals over the crumbling walls. The civet sat on the dark shelf watching us like a ghost, not too acquainted with or scared of humans, living in this forest. It walked slowly and leapt out through the broken window, a long, serpentine cat, nowhere to be found outside moments later. In the coconut trees of our school farm I've spotted civets at night, most of the time by sound—a gnawed-out coconut thudding to the ground, and a couple of times by sight. Yellow eyes catch the beam of a flashlight, fronds rustle and the slender shadow slips out of this world. No amount of search will yield it again.

A large branch, the size of a neem tree, had snapped off and fallen from the baobab. A gaping hole showed its rotting heartwood. The tree, which Geetha says may be a thousand years old, was dying. But even its death would be much longer than my lifetime. I had seen this tree last, deep inside the estate, two decades ago when I was in junior school. About fifteen of my classmates locked fingers and just about managed to group-hug the trunk. There is something about childhood perception which dilates experience. There's surely a general relativity of the mind, of how wonder and attention warp space-time. Seen through child eyes, spaces feel vast, trees reach clouds and events imprint far too deeply, mixing into other parts of the mind's deep sea, then bubbling up unpredictably, a day or a decade later. In my memory this baobab was three times thicker and much wider on top than when I saw it today with adult eyes.

Along the paths, tinospora vines (*seendhil kodi* in Tamil) rained down from all the tamarind trees. We passed a medium-sized banyan—after rushing through a multi-canopy roost of a few thousand raucous flying foxes. This tree according to lore was in the path of a road being laid, and its spirit is supposed to have rebelled against being cut down. It reached out to Bishop Leadbeater and former president Jinarajadasa in the 1930s, as recorded in the society's journal.[19] Both men, theosophists and occultists, narrate how the tree's consciousness was so strong and

[19] C. Jinarajadasa, 'The Two Banyan Trees', *The Theosophist*, September 1931.

evolved that it seemed to warn and rebel and speak. They call it among the most intelligent of trees. A cutting was taken from the tree and given a different place, and they felt it was appeased. From a hollow in one of its branches two spotted owlets watched us wide-eyed as I listened to Geetha telling this story. I watched my animist mind as I listened—which is usually sceptical of human-centric occultism but which made an exception about the aliveness/consciousness of trees.

There's a small hall called the Garden of Meditation near the Great Adyar Banyan, in which I might teach students soon. This is special to me. To the east is the Great Banyan and a seat from which Krishnamurti gave his first speech. To the west is the Olcott bungalow where Montessori taught during her stay in India. It is said that both these teachers needed the blessing of the Great Banyan.

Kesavaram Dam
22 July 2022

If there is just one piece of information I must give about Chennai, and nothing else, which would let an intelligent researcher weave most of its history—ecological, sociopolitical and so on—what should it be? I would tell them of the land's gradient. The land slopes here from around north and northwest to south and southeast. The invisible but powerful agency of the tectonic platform we stand on grips my imagination today as I stand at the Kesavaram dam after tracking the Cooum River from its estuary at the end of Marina Beach, over the Napier Bridge painted in black and white checks, to its very beginning as a branch of the Kotralayar River at Kesavaram. A group of us from the Chennai Climate Action Group has travelled the whole length of Cooum with historian Venkatesh to curate a water tour along this river which had to be killed for the city, as it stands now, to live.

The slope's agency keeps showing itself as we travel counter-course. Rivers naturally flow, branch and spread over the slant set by land. Surface geology is influenced by land gradient again—most of Chennai city is river alluvium, clay and coastal deposits. Land use, occupations and vegetation follow. Land had already loaded the dice in certain directions. There have been many battles fought in this province. Between the British and the French, two between Tipu Sultan and the British, the Pallavas and the Chalukyas, the Cholas and the Rashtrakutas—who fought their war in Thakkolam, a battlefield now visible as trees, grassland and paddy fields from over the dam where Cooum splits from Kotralayar. In this region armies which attacked from the north, from higher ground, almost always won and took over the region. Elevation and vision were a huge advantage for Tipu, who perhaps put up the most fearsome resistance to the British, inspiring struggle in other parts of the country. My language was determined by land slope. I speak Tamil instead of Sanskrit, Prakrit or Telugu because of it.

Rivers and river basins are flyways and migratory corridors for butterflies to migrate from over the hills to the plains and back as the two monsoons, southwest and northeast, alternate between these regions. The air rings with the calling of pied cuckoos. Wandering gliders float in closed alphabetical loops—Ds, Ps, Os—mating over the gold sand and streams of water so clean and brimming with fish and dragonflies that the name Cooum is confusing to use here. Dancing dropwing dragonflies clutch the tips of vetiver grass, wings raised, swinging side to side, balancing on an impossibly small perch.

The 2-kilometre-long southern bund at the divergence of these two rivers which gave Chennai its identity and history is also a dense butterfly corridor. The plant community here is a river-grown, insect-grown massive butterfly garden. Hopbush (*Dodonaea viscosa*) and Indian cadaba (*Cadaba fruticosa*) grow here in numbers I haven't seen anywhere else. They are extraordinary nectar plants. Limes and Mormons are laying eggs on the hopbush, and orange tips and crimson tips—creatures you barely see in the city—on the cadaba. The treebine climber (*Cissus vitiginea*) on the bush-beeches has dozens of bouncing common leopards and red net-winged beetles nibbling at their flowers. Lantana is pushing its way in too. Down on the ground for small butterflies and insects—the line blues, grass jewels and also the felt-topped nomia bees (*Nomia thoracica*)—there bloom justicia, devil's thorn and indigofera, among others. There is heliotropium and crotalaria for the crows and tigers. Just by the bund dragonflies are beginning to swarm and move downriver—wandering gliders, marsh trotters and picture wings. It is a rare insect haven. Two lemon pansies are rapidly circling each other, first near the ground then slowly together, ascending 15 feet into air and spin-wheeling there.

Sheer butterfly numbers and species lift up our energies after the long journey. Nikkitha teaches Prashanth, Vishvaja and Raju butterfly names. The air is clean, the river blue and clear. Its bed has large, smooth pebbles and shy yellow-wattled lapwings. Cooum within the city is as much a pejorative, synonymous with filth, just as it's a river's name. But this is the river before it enters the city. Cicadas perch on the branches of prosopis bushes and buzz loudly.

At Kesavaram the Cooum is hacked off right at its birthplace. As soon as it branches from the Kotralayar, the Kesavaram dam, built in 1947, cuts it off, diverting water to the Poondi reservoir and other water sources. Then, about 25 kilometres downstream, at the Koratur dam at Puduchatram, built in 1878 by the British, the water is diverted to the Chembarambakkam reservoir among other water sources later along the watershed of Chennai.

Till then, Cooum is clean and thriving with bird and insect life. But past Puduchatram we see the water-guzzling bottling plants of Coca-Cola and other companies sap the river further. Then further downstream around Poonamallee, universities, apartments, transmission towers and hospitals are built right upon the river, making its width just over 10–15 feet wide in some places. From here onwards sewage outfalls are all along the river up to the sea.

Once it enters a city, any river becomes a black, toxic, foul-smelling trickle. Cities grow along riversides, then slowly forget their ecological, hydrological genesis. Later they slowly collapse under their own weight—unless perhaps there is a powerful re-invocation of what birthed it in the first place. A river.

Section 6

INTERTIDAL
MEDITATION

Urur Kuppam Beach

Thin moon in the noon sky. I walk to the ocean before an art activity with the youth at Urur Kuppam, just as one would visit a friend having come to the neighbourhood. Low tide has left an intertidal space at least 20 metres wide. I walk within it towards the Adyar estuary following a single sanderling scuffling along the beach searching for lugworms. It shuffles up and down the shore while waves wash in and out. The bird's tiny footprints are drawing waves on the sand. I look back at my own path—it too is wave-like, curving upshore when water surges in and back down when it recedes.

In my mind, a recent conversation with a friend on the new queer glossary in Tamil surges to the fore, invoked by the metaphoric power of the land–sea junction I am walking on being made and remade. The sanderling bobs its whole body and looks back. In Tamil, the word for queer is *paal pudhu*,[1] translating literally to 'gender new'. A word fully open like the low-tide shore. A non-binary person is *eermaraikku apparpattavar*—meaning somebody 'beyond' binaries. Words more in celebration than their English counterparts, and those which can describe and have a depth beyond people and sexualities. The many mergings of non-binary meanings the coast shows me open a window into my own many intertidal/queer spaces within, washing away several opposites, binaries. I remember that in all the living world and especially in the ocean, most species exist in a sexual intertidal zone—hermaphroditism—among other ways. In *Evolution's Rainbow*, Joan Roughgarden writes, 'The most common body form among plants and in perhaps half of the animal kingdom is for an individual to be both male and female at the same time,

---

[1] Notification Regarding the Glossary for Addressing the LGBTQIA+ Community, G.O. Ms. No. 52, Social Welfare and Women Empowerment [SW-3(1)], 20 August 2022.

or at different times during its life.'[2] That is just one example. I come back home inspired and write an intertidal meditation.

(Intertidal meditation to be read out slowly on a safe place at a beach, preferably at low tide. With sufficient pauses between sentences/phrases and longer pauses between paragraphs, giving enough time for inward searching and visualizing. In the end, a sharing of what different people recalled, visualized, felt can be invited within a safe group.)

Face the ocean and close your eyes. Bring your awareness to the sound of the waves, where the ocean meets and merges with land. Let your awareness stretch across this intertidal space, from horizon to horizon. Listen closely to the aliveness of water and anything else you can hear—birds, sounds from the sand, the different textures of waves rising, crashing and pulling back in. Listen to the crashing of larger waves, smaller waves and those which just gently slide in.

Gently bring your attention to another intertidal space, the entire surface of your skin, the junction where your body meets, merges with the atmosphere. Feel the embrace of air over you which keeps your body intact and in shape with its pressure. Feel the gentle hug of the atmosphere from the hair over your head, down to your face. When you reach your nose, notice the tides of your breath, where the air rushes in like a creek and keeps your body alive. Feel the intertidal space in your lungs. Then feel the air–body junction down your neck ... chest ... stomach ... then slowly down your legs.

Softly notice your awareness of yourself—a vast intertidal space between your mind and body. Notice its tides, movements, blurring the distinctions between what is mind and what is body, as coast blurs what is land and sea.

[2] Joan Roughgarden, *Evolution's Rainbow* (University of California Press, 2009).

Now, let the intertidal's wave sound, aliveness, energy flow into you. Explore other intertidal spaces in you which dissolve binaries. Know that the logical part of the mind, which has evolved fairly recently in time, uses the filters of binaries, of opposites, to make sense of the world in simplified terms. Let that nascent part of the mind rest. Reach into a deeper, more ancient part of yourself which observes without judgement, without classifying, which just observes keenly, curiously, even the simplest thing. A place which lets you experience what is beyond opposites and binaries, which lets you see in vivid complexity, richness and queerness—the world and yourself.

Enter yourself now to explore your intertidal spaces. Search your body, your memories, your experiences and feelings where one or more of the following binaries dissolve.

mind–body, male–female, land–water
mind–body, male–female, land–water

Bring to mind one intertidal space or experience within you which dissolves one or more of these binaries. Give it your full attention, to the space it fills in you, to the feelings it evokes. Watch this for a minute.

Now enter yourself again, search your body, memories, experiences, feelings where one or more of the following binaries dissolve.

good–bad, joy–grief, pain–pleasure
good–bad, joy–grief, pain–pleasure

Bring to mind one intertidal space or experience within you which dissolves one or more of these binaries. Give it your full attention, to the space it fills in you, to the feelings it evokes. Watch this for a minute.

Now enter yourself one more time, search your body, memories, experiences, feelings where one or more of the following binaries dissolve.

living–non-living, inner–outer, I–other
living–non-living, inner–outer, I–other

Bring to mind one intertidal space or experience within you which dissolves one or more of these binaries. Give it your full attention, to the space it fills in you, to the feelings it evokes. Watch this for a few moments.

Now gently move your awareness to the embrace of the atmosphere over your entire body. Feel fully this intertidal space, this junction between you and the atmosphere.

Now gently move your awareness to the sound of the waves, of the ocean's intertidal zone. Let your attention stretch across this space, from horizon to horizon. Honour and offer thanks to this intertidal space, and also all the intertidal and queer spaces within you, enriching and deepening you, completing you.

Ennore Creek
8 August 2022

A young kadal shrimp hops on the water with three tail flicks and lands on the dredged banks of the creek under the bridge. Perhaps it is fleeing something pursuing it underwater. I hold it up to show people. It curls its hair-like antennae, much longer than its body, slaps its tail on my palm like a spring and lands back into the water, not swimming away but letting itself sink to the bottom amongst the girdled horn snails. A reef egret stands on a pillar's plinth, shrinking into a thin slab of shade under the coal conveyor belt bridge.

Deep sediments of fossilized oyster reefs lay around in the dredged mud. Two new shrimp varieties have been identified in these waters shadowed by the towers of two thermal power stations. The prawn-catchers call them: *Melpu* era (floating shrimp) floating up on their own from the muddy creek bottom from thermal shock or hypoxia, caught in the hot-water effluent discharge; and *Sambal* era (ash shrimp), speckled grey with a carapace rough to the touch, greasy on fingers when peeling and deveining, from the coal fly ash accumulated in its guts.

Most prawn and shrimp spawn in the sea come to coastal wetlands as postlarvae and grow there, and then return to marine waters when they are almost adults. Ennore is among the most toxic and treacherous waters they'll enter, a brackish industrial Mirkwood where the Kotralayar River meets the ocean. The tide is meraal now, or right past high tide when the water is at a standstill at the creek before beginning to recede seaward. A time when nets are cast across small channels and gradients to catch prawns and fish being carried down by the current.

Our bus goes further north on the port road parallel to the Buckingham Canal on the west and the stone compound of the Kamarajar Port on the east. We stop at Nettukuppam where the wall has an alcove. A metal gate hosts a yellow poster of the deity Ellaiamman (boundary goddess) and the dates of all the new moons this year. Most coastal villages have an Ellaiamman temple to guard the village. But in 2001 Nettukuppam's temple and its lands were taken over by the port authority and walled off. Villagers protested, but the port refused access. Then the people broke down the stone compound one new moon, went in and did their rituals. The shaken port authority has now built them a gate to enter once a month and see their goddess.

I chat with Bhavani Raman during the journey. She has researched 'documentation' as a form of discrimination during the British Raj—how paperwork was a primary tool of dispossession. This continues to be an eviction strategy for the state. She now investigates the lives of salt workers in northern Tamil Nadu, their relationship with coastal wetlands and their struggles with the industries and authorities. Something she says stays with me: 'The binary between land and water permeates our thinking.' This place we are travelling through, the backwaters of Kotralayar, seems to resound to that statement, where there is hardly a habitat which is entirely land or entirely water. We only enter intersectional spaces here, queer ecologies. Tides are the main queering entity; then the turning seasons, creatures, occupations and, finally, language, reflect this. Ennore is full of land–water binary-breaking words in daily, hourly speaking— seru (slush, especially of dark organic sediments), *sadhuppu* (marsh), *paraval* (flood plains), *uppalam* (salt pans), *uppangkazhi* (salt marsh), *eer nilam* (tidal flats) and so on. I think about what happens to our experience of place when land and water are treated fictionally separately, like female and male, nature and culture. What realities are then easily invisibilized and erased?

Over Kattupalli Island millions of wandering gliders congregate over the prosopis bushes and the shell heaps outside lime kilns. Green marsh hawk dragonflies are flying into the Karungali estuary against the wind, from over the ocean, where the Kotralayar's second estuary is blocked by a road built by Adani Ports. Nity explains their port expansion project to the group on a map, on which the concrete of the port in yellow liberally colours over river and sea.

I buy *kot* era (tiger prawn) at the Pulicat market, the largest of Indian prawns and famous at the lagoon's seagrass and seaweed beds. I peel this species at home for the first time and prick my fingers on their spiny carapaces, especially near the tail.

Urur Kuppam Beach
20 August 2022

On this morning's shore walk, an enthusiastic girl of six points to the sky, interrupting my ghost crab story, and yells, 'Isn't that a hornet?' Everybody looks at the fat insect swerving and rebelling against the wind. I say that's a carpenter bee, which makes her raise her eyebrows and repeat the name with excitement. The bee flies down and enters a bamboo pole on a docked boat where the Indian flag and a political party's flag are fluttering in the wind. Carpenter bees are beach creatures, as much as ghost crabs or wedge clams. Over the berm they fly around as if in patrol and near the flowers of sea beans and ipomoea. One often sees late-morning courting pairs zip out over the ocean far beyond the breaker zone.

Like other solitary bees, carpenter bees nest in groups of intergenerational sisterhoods in deadwood and bamboo, with woodchip walls to partition their rooms. I've seen them burrow into washed-up furniture, cane chairs and driftwood lying lodged in these sands. I've gotten stung at the centre of my palm while carelessly gripping washed-up branches and unknowingly shutting their burrows, then the bee flies around me agitatedly, chiding my recklessness. In holes made in the poles of fishing boats bee sisters are taken on trips into the ocean at midnight or early morning, as they sleep.

We learn to track the longshore current's direction by looking at the ocean with one eye, through a thin slit between two fingers or palms. Over forty of us do this, child, parent and grandparent, making other beach-goers look at us strangely. Through a narrow slit, longshore flow becomes clear to the mind. It is difficult to discern the flow if the whole ocean is seen in its panoramic expanse. Today it is still faintly thendi—south to north—and the inshore current is hot chocolate with sediment.

I come home and think about how the mind and the coast have been growing more metaphorically interlocked for me. Ocean could be to landmass what the subconscious may be to the conscious mind. Deep, many-dimensional and unfathomable. Much older in genesis. But the conscious, a relatively recent-born tip floating on the psyche's surface, can feed and shape the constituents of the subconscious like rivers their estuaries, making the nearshore waters rich. Both coast and mind have their intertidal zones, junction spaces where tides fluctuate between the surface and the deep. Dreamy spaces where senses are blurry.

A common phenomenon in all dreaming is their lateral movement relative to lived experience. Elements, emotions, events from different timelines and life stages mix together, jumping from context to context, moving perpendicular to the currents of reason and wakeful reality. They connect sideways the different islands of experiences like longshore currents. By the high-tide line today lay floral decorations—tuberose clumps and garlands of asparagus—from the Madras Day celebrations, carried aslant to this dockyard, lying on the intertidal like strange dream objects. When the northeast winds begin and the longshores reverse, you can find on Urur Kuppam Beach wave-smoothened pebbles of anthracite coal carried down from the thermal power stations at the Ennore Creek.

Then there is the sheer effectual force and life-altering energy which can only arise from the deep sea of the mind. The coastal consciousness can be utterly washed away, but it can also grow to buffer, regulate and thrive on this energy, making resilience like mangrove roots or sand-dune slopes. When repressed, chronically stressed or traumatized, the subconscious rises like sea levels and storms and thrashes landmasses, like climate change in the body. It demands restorative balance and respect. These metaphoric imaginations emerge from co-thinking with the coast. I wonder if our mind has evolved mimicking the biosphere's structure. The wood-wide-web shows how the forest floor is like a mind as well. Is our psyche's ontogeny an imitation of the larger ecology it is entangled in?

Not surprisingly, wayfaring the mind's vastness is as challenging as journeying through the ocean. For some of us both of these embarkings are the same. Inner and outer merge, and one walks on the shores of both.

Chennai
5 September 2022

In the last week of this school term I went on two trips to the farm with groups of older elementary children. We learnt to observe, record and accurately draw butterflies on the field. Children roamed the different habitats on the farm and the Vellaputhur Lake with field guides and observation tables. We spent time in the butterfly garden which had host plants like milkweed, oleander, jujube, bel, senna and palash for local species of butterflies. It also had native nectaring plants like premna, ixora and indigo which their seniors had planted three years ago, and many classes have since been maintaining them.

Common emigrant butterflies were laying eggs on the leaf buds of a senna plant. The passion vines growing wildly had pearly eggs of tawny costers in clutches of sixty or more, along with their spiked-sausage caterpillars and Egyptian-mummy pupae. This time of the year when the monsoon peaks in the Western Ghats, hordes of butterflies migrate to the eastern coast. They will breed and perhaps die here. Their offspring will fly back when the monsoon circles back and hits this region. Changing climate has made the monsoon a fickle beast—coming early, coming late, pouring down all at once or not at all. Core samples, 200 metres long, taken at the Bay of Bengal last year show a million years of monsoon data, revealing clearly from plankton fossils that deluges happen when CO_2 concentrations are high. And right now they are higher than ever.[3]

The same time three years ago when we studied butterflies at the farm, we recorded over thirty species. This time during the two trips, most children saw less than twenty. Common jezebel, common wanderer, painted lady, joker and great orange tip were among the species missing. While Karachi and Bengaluru were receiving disastrous floods, here we had the hottest September we have experienced.

[3] Steven C. Clemens et al., 'Remote and Local Drivers of Pleistocene South Asian Summer Monsoon Precipitation: A Test for Future Predictions', *Science Advances*, June 2021.

Earlier this month the Greater Chennai Corporation released Chennai's draft Climate Action Plan for public review and feedback, which showed some dire truths about the future of this highly climate-vulnerable coastal city. In the next five years a sea-level rise of about 7 centimetres is bound to cause 100 metres of the coast to be swallowed by the ocean.[4] This would submerge a massive width of existing beach habitats, nearby residences and artisanal fishing hamlets. Low-lying areas including the Pallikaranai marsh would have sea incursion, turning many freshwater habitats brackish. Groundwater would be severely impacted, 29 per cent of the city would be inundated during floods, and floods and cyclones were bound to increase with the erratic monsoons. During summer 53 per cent of houses dependent on external water sources would be water-stressed. These are five-year predictions. The report's risk assessment includes twenty-five-year and hundred-year predictions as well, which my mind didn't cognize for the moment.

———

A colleague, from a totally different field of teaching, asked me in a planning meeting this week why I waste time talking to children about trees, butterflies and birds. What's the use? How does it matter? Are we just trying to kill time? I listened to his questions closely. It cracked my world like the way an old woman made the hills crack in Arun Kolatkar's eponymous poem. First shock, then pangs of anger that had to be held down. My identity was being grabbed by the throat and shaken, but I was able to witness it without reacting. Then a perceptual portal opened by itself and I asked him (within me, without speaking): Who are you? What shapes your seeing of the world? To encounter those who find

[4] Climate Action Plan Chennai, Greater Chennai Corporation and C-40 Cities, September 2022.

my work irrelevant is commonplace. 'Difference' is a requisite for both life and conflict. I tried to listen to him as I recalled what psychologist Carl Rogers had said about listening: 'Listening requires wearing the other's point of view and feelings as a mental costume, without analysis or judgement.'[5] He also said, 'To listen to oneself is a prerequisite to listening to others.'

In subsequent meetings with friends and colleagues I sat consciously wearing a lens of difference—that each person speaking and listening is utterly different from me and others. This was a gift gotten from paying keen attention to my colleague's contempt for my work, fully justified by his lived reality. I saw listening and paying attention as a disarming act of entering someone's unknown otherness. I try to leave my perception porous during these times.

Painted grasshoppers are fully grown and mating on top of the milkweed plants around my home. I've been watching them since they were nymphs, which emerged from their undersoil egg pods after the first summer rains. They are now 10 centimetres long and gaudy blue, yellow, red and green insects wilding the waysides. I come to the grasshoppers mating in the sun with the same questions: Who are you? What shapes your seeing of the world? The insect pairs move into a leaf's shadow, trying to hide from my gaze.

I notice how attention can be free of magnitude, unlike in the physical world. What you see and feel fills the mind's scale. A moth, microbe and mountain can exchange sizes and prominence in mindspace. Somebody's large is always another's speck. For a while the activities of a pair of grasshoppers become a larger-than-life issue. Cultures clash across species—the adults are

[5] M.M. Owen, 'The Art of Listening', *Aeon Magazine*, May 2022.

mating right before the nymphs, and I reflexively wonder what the small ones think is happening. The males are several centimetres smaller than the females and it appears they like to climb on the females and perch on their backs for long hours, if not days, before mating ensues. And during the mating, the female may feed but the male might not be able to. These insects always prefer the top one-third of the milkweed plant to eat, perch, bask, mate. Smaller ones largely eat the flowers and leaf buds.

My place in the food web and the purview of my senses limits and frames my seeing of the world, especially when seeing across species. Grasshoppers are usually competitors—we eat the same plant species. They have been historical drought-bringers. Human knowledge of them begins with the crops they eat and ways to exterminate them at different stages of their life cycle. Between 2019 and 2021 locust swarms emerged in Africa due to climate-change-induced droughts and flew to India, destroying more than 2,00,000 hectares of crops.[6] A locust is born to a climate-stressed mother grasshopper who lays mutated eggs. 'Pest' is the preferred term for grasshopper. Painted grasshoppers don't swarm, though, and they eat milkweed, a plant which is poisonous to us, yet grown in some places for medicine and for its flowers.

It was once thought and portrayed that all plants do is compete with each other for sunlight. Later our minds grew to see that plants form friendships and collaborate on vast scales underground. It taught us to be wary of simply framed realities. There are only complex, inscrutable and ambivalent relationships in the universe. Today I think from my agrarian culture perspective that painted grasshoppers merely eat milkweed and threaten crops—a simple frame. These plants are also their homes, hiding places, nursery

[6] Kabir Agarwal and Shruti Jain, 'Climate Change Brings the Worst Locust Attack in Decades to India', Wire, May 2020.

grounds, mating spots. What is the complexity and ambivalence I am missing which my eyes cannot see yet:

> Because truths we don't suspect have a hard time
> making themselves felt, as when thirteen species
> of whiptail lizards composed entirely of females
> stay undiscovered due to bias
> against such things existing.[7]

Surely there is something extraordinary to be found about the grasshopper and its wondrous entanglement with the milkweed and other beings. I don't know yet if this is true. The only truth after all is perhaps the act of paying attention.

Chennai
31 October 2022[8]

love / / meaning live in intimate observation
wanting others to know / wanting it purely ...
to expend the heart
...
to transform one's self to the other /
to open cell to cell.
 —Jo Mariner

A black-hooded oriole is hovering and nibbling a half-ripe papaya on the tree and feeding it to its fledgling sitting on a branch. Then it flies away to a gooseberry tree's branch nearby and watches its young one keenly, as if to see if it would begin feeding on its own.

---

[7] From the poem 'Cascade Experiment' by Alice Fulton in her collection *Powers of Congress* (W.W. Norton, 1990).

[8] This passage first appeared in *Vikalp Sangam* as 'Can Biodiversity Be a Political Science Teacher?' in January 2023.

At the end of October four of us nature educators are in the village of Pitchanur in Coimbatore for two full days of activities with the local children. The previous day we had gone birding, done nature-journalling and played a few games in the afternoon. Around fifty children had come, some from Pitchanur and adjacent villages, and others from an Irular tribal settlement located a little outside the village. The fact that caste segregation ran deep in this landscape, even among children, was only slightly apparent to me, an outsider. In how they sat to eat lunch, in how some children did not budge when I tried to make small groups of them (so I let their teacher do it instead). The next day by popular request it was half a day of butterflies. In the morning we went on a long walk in the village outskirts, field guides and observation tables in hand, searching for butterflies. We saw common banded peacock butterflies (*mayil azhagi* in Tamil) mud-puddling on a heap of wet red soil. A southern birdwing (*ponnazhagi*) patrolled over the coconut plantations, bringing much excitement and yelling each time it passed overhead. Along the waysides between plantations and fallow lands, four-rings (*nangu valaiyan*) were plentiful with their slow flutter low over the verge grasses. We also saw the full life cycles of the plain tiger (*vendhaya variyan*) and the lime butterfly (*elumicchai azhagi*) on a calotropis and a lime plant respectively.

Towards the end we sat on the banks of the Kumittipathi River to share our findings, observations and questions, and listen to each other. Some children had spotted over twenty-five species. Later in the day, Sandhya, one of their teachers, shared with us rather movingly that somehow during the activity the children had gradually started interacting, and by midway they were freely talking to one another, helping each other find, identify and observe. This was something she said she struggles to bring about on a daily basis, given the sociopolitical setting of the place.

This occurrence left me thinking for several days after we returned to Chennai. What was in the nature of watching butterflies or birds or trees that was able to erase, if only for a few hours, such deep social segregation? Was it simply that while watching a butterfly together, caste was irrelevant? Or even more significantly, in order to pay attention, to wonder and to raise questions, one had to drop away social constructs, if not bring those constructs also into observation, and be on a humanely equal plane? Something about observing deeply and connecting to other beings could connect human beings too.

A few days ago I was presenting to my school possible ways of integrating climate learning as a core aspect of our education. My colleague Kaveri, a teacher of humanities and political science, shared something deeply intriguing. Her grade 9 children who have been through our 'Farm, Environment and Society' programme since elementary school were far more politically aware, capable of thinking for themselves and active discussants than the other classes she had taught. The nature-based pedagogy we followed in the last few years seemed to have transferred into other areas of life. She felt that 'something about the skill of observing keenly for oneself' extrapolated into their engagement with human society and their thinking about history. I heard Mahesh Rangarajan, a professor of environmental history, say in a talk that 'engagement with biodiversity can help us meet all kinds of otherness'.

What are the ways in which biodiversity is a political science teacher? People have been unearthing numerous transformative political ideas from other species. Alexis Pauline Gumbs finds profound teachings of resistance and ways of shattering capitalism from whales, dolphins and seals.[9] Jean-Paul Gagnon in his three-part essay explores the question of non-human democracy and

[9] See her brilliant book *Undrowned: Black Feminist Lessons from Marine Mammals* (AK Press, 2021).

uses 'interspecies thinking' to draw operative democratic lessons from bees, bonobos, termites and microbes deeply applicable to human society, as well as provoking ideas for what a multispecies democracy might seem like.[10] In the book *Evolution's Rainbow*, Joan Roughgarden tells us ways in which other species, from insects to fish, can teach us to live in a diverse society, especially a gender-diverse one. That nature is profoundly queer, that binary, polarized nature is hard to come by. But at a more simple level for a child, what are the subtle political learnings that happen through a regular practice of observing nature?

I realize, from experience and from reviewing research, the simplest and most formative political value that direct engagement with nature is able to offer is an immersive exposure to 'diversity'. Children learn implicitly that there is never just one voice, one narrative, one story in a profoundly non-binary multispecies world. One need not even highlight this truth as an educator. 'Difference yet coexistence' is the lens through which the living world lets itself be seen. Other beings speak to us subliminally. Observing biodiversity can shift us from the consumer–recipient location that most human beings have been cornered into by capitalist culture. Direct observation makes us active foragers of deeper meanings and purposes—which by itself is politically countercurrent. Gregory Cajete, a Tewa elder and educator, writes: 'Observing how things happen in the natural world is the basis of some of the most ancient and spiritually profound teachings of Indigenous cultures. Nature is the first teacher and model of process. Learning how to see Nature enhances our capacity to see other things.'[11]

[10] Jean-Paul Gagnon, 'Non-human Democracy', *Conversation*, December 2015.

[11] Gregory Cajate, *Look to the Mountain: An Ecology of Indigenous Education* (Kivaki Press, 1993), p. 223.

My teammate and bold young nature educator Nikkitha describes how she and her friends have developed a daily practice of looking—in the waysides, shrubs, grasses—especially for 'what is not easily seen'. This they felt was the beginning of critical thinking which transferred to other areas of their life and to their interactions with people—the perpetual effort to look for the invisible or the invisibilized. Surendhar Boobalan, a friend and fellow nature educator in Puducherry, pointed out to another aspect of equality which emerges when he takes primary class children for birding. That he is no longer able to notice the distinction between studious and unstudious, bright and dull children—which a confined classroom sometimes forces him to.

The political–pedagogical processes one follows as a nature educator are vastly different from traditional classroom instruction where power and spotlight is concentrated on one person—what I've begun calling a 'pedagogy of control'. In a marshland or a park, if a frog or a heron decides to show/teach something else and the learners' energies flow in a different direction other than my own plan, I have learnt to leave space for it—for nature to directly be the teacher—aware of the fact that I am always both educator and learner in that setting, as is everybody else. When the learning space is the real world, the educator has little choice but to drop control and evolve a 'cooperative pedagogy' where power, knowledge and focus is distributed multi-people, multispecies. These multispecies values are already present and practised in several Adivasi cosmologies—the Idu Mishmi, the Santhal, the Jenu Kurubas and the Kattunayakars, to name a few.

Through the Palluyir Trust (*palluyir* in Tamil means biodiversity/multispecies/all of life), and in collaboration with Pudiyador (an organization which works to empower marginalized communities across Chennai through education), we run the Youth Climate Internship, a programme for youth from three climate-vulnerable communities—Urur Olcott Kuppam, Ramapuram and Kakkan Colony. Through the programme we make ten field trips to observe and understand the ecology and sociopolitical landscape of Chennai. We learn advocacy tools and the law; we study other species and habitats in our neighbourhoods and the youth engage people in their locality in walks and activities. Recently, on a cold Sunday morning at Urur Kuppam, we had a half-a-day module on 'questioning'. In the morning we did a 'curiosity map', an exercise to actively strengthen our muscles of wonder and curiosity.

All through the week, blue buttons were washing ashore along the city's coast—a phenomenon which happens two or three times a year, sometimes due to strong landward winds or seismic events and at other times unexplainably. We pondered about the recent blue button beaching, then asked questions about it, covering which, when, what, why, how and who. We made sure we asked questions past what the mind could easily think of and across the threshold of comfort, consciously challenging our capacity to wonder. Then we headed out to the beach to each make a curiosity map of our own.

The winter sun was two fist-spans over the ocean and pleasantly warm. Some fisher boats were coming back, having cast crab nets early in the morning. An olive ridley sea turtle had washed ashore dead, with an impact injury on the bottom right of its shell, possibly from a trawler strike. Among the sixteen of us, we each chose one creature or scene on the coast and exercised our curiosity. We drew and coloured, then made a map of questions, consciously pushing our wonder beyond its zone of comfort. Decorator worms, tower shells, crows, ark shells, ghost crabs, goose barnacles, a sand star and the sea turtle helped us exercise our wonder.

'How do barnacle shells form under the sea?'
'How does a clam make the inside of its shell soft and the outside rough?'
'How far can a turtle see inside water?'
'How does it help a tower snail to be shaped like a screw?'

'What happened to the creatures inside these empty shells?'
'Can turtles dream?'

To wonder, to question as a daily practice of living is a radical political act. They help change age-old, often obsolete, social constructs and myths holding in place structural inequalities and patterns of capitalist existence on the earth. Wonder will keep alive constant reimagination—political, cultural, spiritual—which is perhaps the mark of a sapient species.

Kovalam Basin
1 November 2022

I am on a recce along a very important watershed of Chennai, the Kovalam basin—which is part of the Adyar and Palar river basins—curating a water tour for our youth climate interns.

Stop 1: Tirusulam Quarry
To repeat my friend Maria Faciolince's words, 'cities are inverted mines'. I hold that in mind. Buildings, roads and other structures have an equal volume of earth, water, people and species missing here and/or somewhere else. Most of a city's footprint is outside of it. Somewhere, beaches and riverbeds have gone missing as sand and cement seep towards sky. Somewhere, hills and mountains have gone missing for stone in infrastructure and roads. It is sometimes difficult to imagine this from within the desiccate aesthetics of urban space and see 'inverted mines' in our minds. The quarries at Tirusulam take this difficulty away.

All of us stand peering down into a massive void of several missing hills. Its cavities in the earth are so deep that black kites soar below our feet. These are spent quarries of blue granite— the gravel and basic material in all construction—mined since colonial times. On Google Maps, the eastern edge of one spent quarry is marked 'hidden lake'. They are used as lakes now, where rainwater and excess water from the Chembarambakkam reservoir is diverted. We imagined what these hills may have looked like and how they may have overlooked the land with their streams, clouds and forests. What happens to the land and its climate when its hills are carved out? Further ahead are visible half and two-third hills being razed down by excavators for gravel. As we travel down the dusty road along the southern edge of the void, we see a series of blue-metal processing companies with fleets of lorries (named earth-movers) and machines manufacturing m-sand and p-sand. Here is the dark shadow of the city, the 'inverted mine'.

Stop 2: Pallavaram Hill

We walk up the hill until metal police barriers let us walk no more. Pallavaram holds a history back to the Stone Age. Asia's first Paleolithic stone tools were found here by Robert Bruce Foote. People have been living here for 2 million years on the Adyar River's banks. I try to imagine what their view from here would have been like and what their lives would have been like. We stand at around 120 metres altitude. Two pale grass blue butterflies perch on the tephrosia flowers growing on rock edges, telling us we are no longer at sea level. Bracken ferns grow from the rock faces. Ribbons of six or more common emigrant butterflies loop and chase each other through the branches of Siamese cassia trees as if they are strung together. Up this hill is a temple, open for a few weeks a year and during festivals. The sacrality it has cast upon the hill protects it from quarrying companies who've gouged out nearly all of its siblings. From the temple we look down east at the expanse of South Chennai through our binoculars and try to identify the large buildings, lakes and localities from this kite's-eye view. It is impossible to visualize what the view of the Paleolithic people might have been like.

The Chennai airport sprawls right at the foot of the hill and from here it looks like half the city. The runway is a sterile and massive greenfield with metal birds parked, landing, taking off. This airport is 1,300 acres on the Kovalam and Adyar basins. It courted further disaster by building a new 2,925-metre runway in 2011 right by the Adyar, when various people warned how that puts the rest of the city at risk of flooding. Krupa Ge points this out in her book *Rivers Remember*. We think about the second greenfield airport proposed for Chennai at Parandur, against which farmers of the villages upon which it is proposed to be built are protesting. This falls in the Palar basin spanning over 4,197 acres. Almost 90 per cent of the area consists of water bodies, agricultural wetlands and grazing lands. If the airport construction goes through, the lakes it will fill over and erase are the Edayarpakkam Lake, Valathur Lake, Akkamapuram Lake, Ekanapuram Lake, Kannanthangal Lake, Mahadevimangalam Lake, Singilpadi Lake, Nelvoy Lake, Attuputhur Lake, Vel Lake, Parandhur Lake and Thodur Lake, and numerous acres of pasture and grazing lands fed by these wetlands.

Stop 3: Periya Eri

Down state highway 109 we stop at Periya eri towards which slopes down the watershed of the dismembered hills. The highway bisects it and goes further to cleave the Keelkattalai Lake and the Narayanapuram Lake. Small squares of paddy still exist near Periya eri but prosopis bushes and newly sprung apartments block them from view. We take a paved stone path constructed for a nearby apartment and walk along the lake. Two pairs of coots dabble where the water hyacinth carpet ends. A few hundred barn swallows swim in the thick mist.

A couple of men are wading chin-deep in the waterbody not 10 feet away from the coots. They have three layers of mosquito nets around large, rectangular frames and long plastic bags held between their teeth. When one of them comes out we ask what they are catching in the water. Selvaraj, a man in his late twenties, tells us they are ornamental fish farmers from Kolathur 30 kilometres away. The most important food for the fingerlings of the fish they grow is daphnia or the water flea—a tiny white, mustard-sized aquatic crustacean when fully grown, but zooplanktonic during its larval stage. His descriptions and knowledge of this near-microscopic creature are deeply embodied, based on his years of practice. He says, if you touch the water the temperature shouldn't be too warm or too cold. Right now it is a bit cold after the rains and not conducive for daphnia. When you skim the water's surface it should be mildly green, not clear but also not dark green. That is the *paasi* (cyanobacteria) which daphnia like to live in and eat. The water mustn't be pure. There should be some, but not too much, sewage mixed. Daphnia prefers to be at the junction where dirty and good water meet, at its own Goldilocks zone.

The men had to come to catch the water flea close to dawn, for if it gets too warm they settle at the bottom. More fishers arrive on the path, undress and get into the cold, foggy water. The coots keep a constant distance from them but wade in tandem with their movements. Possibly the birds too know where to find daphnia and are seeking them out with the men.

Selvaraj says the fish hatchlings like to eat 'nice', their word for the zooplanktonic daphnia larvae. I see the water within his plastic bag and the sheer diversity of benthic creatures from the lakebed—backswimmers, stoneflies, aquatic beetle larvae, water scorpions, and here and there a flickering of minute daphnia-like shiny oil beads.

These livelihoods and webs of lives continue quietly, invisibly on the edge of the highway. But it is almost always the invisible that is the backbone of everything conspicuous. A healthy learning culture always asks: What is it that I don't see? A culture which constantly invisibilizes is one which forsakes its spine and will crack under its own weight.

Barn swallows above fly in random paths, pulling wheelies, stretching wings and swerving, diving, gliding in the fog. Their numbers are high, especially during late dawn. They are possibly feeding on aeroplankton being drifted up by evaporating water. Swallows are aerial filter-feeders, and it's likely there are daphnia larvae in the atmosphere too during this time. The tiny creature has attracted a diverse assemblage of species to the lake.

The fish breeders tell us that Pallavaram Periya eri is one of the last places in Greater Chennai where they still find large populations of daphnia. They are also present in the Shollinganallaur canal and the Manali wetlands, the two other places where catching them is somewhat economical. They were once present in the Perungudi and Velachery lakes, but seem to have vanished after too much dredging and renovation work was carried out there. Water fleas are sensitive to the many balances in their habitat.

Two of the daphnia-catchers are wading ashore, dragging their long bag of sludgy water and their nets. A pair of coots swiftly swims to where they stand and forage in the muddy water they have stirred.

Stop 4: Okkiyam Madavu

We drive over a small bump on the old Mahabalipuram Road and a short bridge curves above a stream flowing underneath. We park the vehicle on the service road and walk along the bridge's edge. Okkiyam madavu (*madavu* is another word for canal in Tamil) drains the Pallikaranai marshland, which is a large part of the Kovalam basin, into the Buckingham Canal, which travels 13 kilometres south before spreading into coastal backwaters at Kelambakkam and reaching the sea at the Muttukadu/Kovalam Creek. A large swarm of shoreflies, little sewage-loving insects, settle in circles over the flowing water. The heat and stench rising from the stream hit our faces with gaseous force as we look down at the water. It is greenish-black with decomposition erupting in it, carrying the discharge from the surrounding IT complexes, hospitals, apartments and shops. From this bridge one can get a sense of the whole watershed.

Far to the west are the Vandalur and Pallavaram hills sloping into Pallikaranai. A wetland has a catchment and a reservoir, just as a washbasin has a catchment and a drain at the centre, or a funnel has a catchment and a pipe. A wetland's catchment slows water down, lets it percolate into the ground and supports life and livelihoods from the fertile soil made by collecting sediment. It is crucial for the wetland's health. All wild places, sanctuaries and forests—the most protected areas—are in the catchment of wetlands. From the bridge one can see that the entire 80 square kilometre catchment area of Pallikaranai was built upon, leaving behind a reservoir spread over barely 12 square kilometres. These paved, impervious spaces within the catchment get flooded each year because water doesn't read the plans of the Chennai Metropolitan Development Authority.

Stop 5: Sholinganallur/Perumbakkam Marshland
A flock of whiskered terns flies over and follows around a shoal of about fifteen stinging catfish slithering in the shallow part of this wetland cut away from Pallikaranai by the ELCOT special economic zone. The fish are too big for these small terns, but the birds just seem to be messing around with the fish, whose long spines, like whiskers, stick out of the water like a carpet of hair. Terns plummet and mock-strike the fish. The fish are churning together and seem to be foraging for something else. Two little cormorants fly away as the school approaches them. On a small grassy islet six spot-billed ducks sit with their heads inside their wings. A pair of blue-tailed bee-eaters hunt ditch jewel dragonflies over the marsh and come back to a thorn bush to feed. I've been seeing these migratory bee-eaters use this and only this plant as their perch at this habitat each year.

Less than 2 kilometres southwest of this marsh are the resettled colonies of people who originally lived on the banks of the Cooum and Adyar rivers, as part of Chennai's river restoration plans. They were made to move from marshland, many of them during the 2015 floods, into an even more flood-prone area. But can it be called flooding if we are made to sit within a waterbody and feel the water rise?

At the end of the Sholinganallur marsh I search for the peregrine falcon on the pylon towers built over it—a black cowl gazing down into mist, keenly studying the tiny movements of ducks and egrets, its grip seeming strong enough to bend the steel crossbar. No luck today. But on the third tower we spot an osprey and feel elated. Two crows on the bars perpendicular to it sit cautiously, unsure whether to mob it or not, assessing its temperament with nodding and crouching heads.

Urur Kuppam Beach
5 November 2022

Seen from the far end of the beach, the strong vanni (north–south current) makes the ocean look like the Brahmaputra. Nearshore waters are chocolate-brown with the sediments flushed by many days of good monsoon rains. But fisherfolk say that the undertow is strong now, going thendi (south–north), making fishing challenging. Locally this is called *iruva,* when the surface current and the bottom current flow in opposite directions. I found it mind-like in its ambivalence and counter-flow.

Bottom-set nets like crab, squid and prawn are difficult to cast now, given that the boat will float in one direction and the net in another. Floating nets for sardines and anchovies work better. I've learnt from fisher friends at Urur Kuppam that ocean currents don't diffuse easily. They retain their character for long distances and time. They almost have a skin around them. Rachel Carson called ocean currents the global thermostat, making up for the unequal distribution of heat on earth. She writes in *The Sea Around Us* that it is possible to track a mass of warm water originating in the southern hemisphere and moving north for 7,000 miles and 1.5 years, till it merges with the rest of the ocean and becomes unrecognizable. Now when I think of water currents, I imagine them sometimes like great eels gliding through the waters, with a distinct will and life.

I was at the beach with my friends to shortly participate in a rally against the introduction of genetically modified mustard in India.[12] This is being pushed by the central government and Bayer, the notorious multinational biotech firm which has bought Monsanto, a company Rachel Carson fought against to ban the use of lethal pesticides like DDT. Threatened by her work, Monsanto made a parody of her paradigm-changing book *Silent Spring*[13] by publishing 5,000 copies of a brochure named 'The Desolate Year', depicting fictitious scenarios of insects overrunning the planet in the absence of pesticides and causing famines.[14] Bayer now pushes GM mustard, a seed made in Indian labs, alongside its own glyphosate, a deadly herbicide which will kill all other plants in the field except their gene-altered mustard.

Placard in hand, I stood facing the ocean today with questions. Last night I read psychologist Donald Hoffman's theories and equations[15] on how consciousness is the fundamental quality of reality, and its constant evolution and manifestations are the many forms of life, matter and particles (not the other way around)—all of them an equal expression of this fundamental commons. Yes, in Hoffman's formulation, consciousness is a

---

[12] *The Hindu*, 'Activists to Strengthen Protests against GM Mustard', 4 November 2022.

[13] Rachel Carson, *Silent Spring* (Houghton Mifflin, 1962).

[14] *Monsanto Magazine*, 'The Desolate Year', October 1962, pp. 4–9.

[15] Donald Hoffman, *The Case against Reality: Why Evolution Hid the Truth from Our Eyes* (W.W. Norton, 2019).

universal commons—an anticapitalism at the very fabric of reality. Of course, many philosophies, dominant and especially indigenous, have had their cosmologies rooted in this commons in their own unique worldviews for ages. This also implies that a bee, a river, a gecko and a human have equally valid and valuable perceptual fields, none of them more special than the other, thus toppling the Aristotelian hierarchy of living beings with 'Man' at the top.

The ocean spoke immediate meaning into this and currents swirled in my mind, volts running down my back, as if something massive and mystical had just shown a vestige of itself and disappeared. Water is a great metaphor for consciousness. It is almost consciousness itself in the way it inextricably unifies and manifests. Post-human feminist Astrida Neimanis says water lends us 'transcorporeality'. She writes, 'While a dominant ... western ... ocularcentrism suggests that bodies are (and should be separate) and discrete, water is in fact digging stealth channels through us all. This connection may be immediate and direct, or delayed and removed, but it nonetheless reveals itself as a thread of interpretation and commonality that facilitates ... an embodied hydrocommons.'[16] The premise of pesticide is based on a view of separateness, a dominant but false view. But the water cycle will make sure through its transcorporeal world-making that glyphosate enters all our cells too, along with the cells of the fish and the osprey, the bee and the worm. It will show us that the individual is a cultural construct that water doesn't agree with.

All water is one contiguous oceanic mass. Water can manifest as river and as ocean currents, but also as whale and fish and sea turtle, breathing and breeding and migrating within itself. It can also manifest as a few hundred people protesting poison

[16] Astrida Neimanis, 'Bodies of Water, Human Rights and the Hydrocommons', *TOPIA*, November 2009.

in our food, as me holding questions to it and as mangroves on the Adyar estuary nearby, in our own ways probing and declaring its nature, our nature, probing and declaring itself/ ourselves—all multiple forms of agency water can invent. If one can see this 'hydrocommons', then any waterbody can directly clarify the commons of consciousness.

We are amniotic. Our skins are membranes, semi-permeable. We are ultimately soluble. Physically. Spiritually. Hoffman invokes Godel's incompleteness theorem as a key component in his work—that is, put simply, there is always more to explore, more to discover, so this fabric of the universe will keep learning, growing, cycling, exploring. 'Truth' then becomes really a state of search. It also means that this truth of consciousness can never be caught completely in any theory or ideology. It will slip out of any attempt to confine it, box it in one way or the other. It will allow you to hold it gently like seawater in your cupped hands and peer into it to see your own reflection. Grasp it and it is gone, leaving behind only a sticky wetness on your palms.

Urur Kuppam Beach
18 November 2022

'Everything is sacred when you take time to notice.'
— J.J. Heller

I stand on the shoreline and count crimson rose butterflies. In spans of a minute on my phone's stopwatch: eight a minute, twelve a minute, five a minute. They are all flying south in straight, tail-streaming paths over the beach and ocean, between eye level and 15 feet high. Ten a minute. Mostly singly but sometimes in pairs and trios closely behind one another. Many flutter down and take nectaring pit stops at the buttercup flowerbeds (*Turnera subulata*) growing all along the backshore near the stone walls of the Theosophical Society.

Seven a minute. A cumulonimbus shaped like a subway burger is emptying a long curtain of rain. In the morning light reflected off the orange sea it looks like water is pouring upwards into the cloud. To the west, a rainbow keeps flickering over the five-star hotels. Mild north wind makes the flaps of my black rain poncho flutter. I've been wearing it often these days, when it has been raining and the light is different. And I've been telling the kids that I'm doing a 'Basics of Dementing' course in Azkaban. I find a few butterflies along the beach pressed to the intertidal sand, their wind-worn wings all veins and leached of colour.

In brief shows of the sun the crimson rose numbers surge to over fifteen a minute. All of them are dispersing towards Sri Lanka with the northeast monsoon's arrival. Soon after the roses are seen migrating, the longshore currents reverse near the coast from thendi (south to north) to vanni (north to south). Each year crimson roses are observed in thousands passing over Dhanushkodi in Ramanathapuram to cross the ocean over Adam's Bridge to Sri Lanka's Thalaimannar, where the ocean's width is the narrowest, about 24 kilometres, between the two landmasses. In some years their numbers are much greater than in other years. In February 2022, naturalists Paulmathi and Vinod saw them migrating in tens of thousands and the spectacle was widely reported in both countries.

This migration is a great mystery. The migration of other butterflies happens between the western and the eastern coasts of the Indian peninsula. Blue tigers, common crows, striped tigers and other butterflies come to the east coast during the southwest monsoon and breed. And it is most likely that it is their offspring which then migrate back to the Western Ghats during the northeast monsoon. Why do the crimson roses alone go to Sri Lanka? They use the coast as their flyway. This behaviour has been described by British naturalists Walter Ormiston (in 1924) and L.G. Ollyet (in 1942). But they have never been observed coming back. I ask Isaac Kehimkar, known as the butterfly man of India for his lifelong work on these creatures. He too says it is not known whether there is a return migration and that we must find out about it. Fishermen from Urur Kuppam tell me that they are seeing crimson roses flying south or floating in the water several kilometres inside the ocean as well.

I ask Krushnamegh Kunte, who is the principal investigator at the biodiversity lab at the National Centre for Biological Sciences and also among the foremost Indian butterfly biologists. He tells me, 'The yearly migrations (for instance, milkweed butterflies) are driven by seasonally recurring climatic events (cold winters, rainy season), whereas dispersals as seen in emigrants are driven by resource exhaustion. Although the crimson rose movements have been described as migrations, I suspect that they are really dispersals.' Which might mean that these butterflies, once they breed in southern India, fly away in the latter part of their life to Sri Lanka, to spend their next few months/weeks. Each generation every year makes this pilgrimage.

Eight a minute. All through the second and third week of October, blue buttons have been washing up on the Chennai coast. At the beginning of the month there was an earthquake in Myanmar. This is the third time in the last few years that I've recorded a blue button beaching after a seismic event somewhere, either far or near. Many examples abound of jellyfish too moving away en masse after or before earthquakes and then beaching. On 10 December 1999 there was a power outage in Luzon, Philippines, when fifty truckloads of jellyfish moved shoreward and got sucked into the Sual power plant's cooling system. This was shortly after a massive earthquake hit Zambales a 100 kilometres away.

Insects disperse usually to leave old and over-populated areas, to find food and to escape harsh weather. During the adolescence of humans and other animals—an intertidal period during our growth—there is a phase also identified as 'dispersal'. There are an urge to move away from the natal environment and to discover one's independence. Risk-taking and novelty-seeking are associated with this phase. This dispersal is not always just a physical moving away but can be an ideological and cultural departure, to 'escape harsh weather'. This is a process hardwired to ensure species survival, and I think we are seeing this happen for good reason, world over. Young people are moving away from ideologically barren lands and obsolete values. There is nowhere else to move to physically. But there is a crimson-rose-like movement, without a return migration, of heeding the environment and of attending to the climate's calling, of moving away from the old political, religious and cultural values that have proved to be ecologically disastrous. There is a reimagining of new values, ones to live by here and now. Twelve a minute. I put my full hope in this.

ACKNOWLEDGEMENTS

My first gratitude to my mother and the banyan tree—for my childhood. Then to all my teachers at The School KFI, its campus, the Indian pitta, the orange-headed thrush, the forest wagtail and the grey mongoose. To all the educator-learners and learner-educators at Pathashaala KFI, its campus and its wild surroundings, the blackbucks, the golden jackals, the migrating butterflies—common emigrant, common crow, blue tiger, common albatross, crimson rose, among others. Then to all the teachers and children of Abacus Montessori School and Songlines Farm School, and the Indian rollers, carpenter bees, robberflies and black-winged kites.

In many stories it is sea turtles who first taught humans how to navigate the sea and wayfare its vastness. They were among our most important mentors, our lodestars. I thank my most important mentors—the sea turtles in my life—who have lifted me by their work and lofted me by their presence, and guided me in rough seas: G. Gautama, Vijay Kumar, Pradip Krishen, Nityanand Jayaraman, Ashish Kothari, Robert Macfarlane, Shobha Menon, Arun Venkataraman, Kamini Sundaram, S. Balakrishnan, S. Palayam, Seetha Ananthasivan and Mahesh Rangarajan, along with olive ridley sea turtles which come every winter to nest on the beaches of Chennai.

Of them, the writer from whom I've learnt the most about writing and of being a good human being is Robert Macfarlane. Special love to him, and to the Goldfinches of Kinnaird Way, Peregrines of the Chalk pits, Gatekeeper butterflies of Orford

Ness, the grand old Oriental Plane tree of Cambridge and the Nine-wells springs.

Gratitude and deep-love to my partner Rega for being my bedrock during the final stages of editing this book and then ever since, along with the Flying foxes of Adyar estuary, Wandering gliders over Kattupalli island, Green-drake mayflies of Cherwell river, the Great Copperbeech at Oxford, and loitering Paper-wasps of Delhi's Green park.

I am grateful to the blue buttons which beached and caused me to be life-changingly curious about the coast and ocean, and the sand plovers, ipomoea creepers, Caspian terns and ghost crabs of Urur Kuppam Beach. Then the many citywide webs which helped me find roots and grow in Chennai—Madras Naturalists' Society, Chennai Climate Action Group, Vikalp Sangam, Nizhal, Sanctuary Asia, Young Naturalists Network, Fridays for Future India, Let India Breathe and There Is No Earth B, among others. My whole team at Palluyir Trust, the extraordinary young nature-educators with whom I work on a daily basis—Jomi, Claudia, Gowtham, Aravind, Kanishka, Hema, Karunya, Prem and Sridevi. We create a range of curricula and educational resources for this landscape and waterscape of which we are a part. And the Ocean 6 team—Vikas, Anooja, Nanditha, Aswathi, Rohith—with whom for two years I have worked to document North Tamil Nadu's coastal biodiversity areas and the threats they face.

I owe thanks to the black kites and pelicans I see every day of the Southwest monsoon, flying south to the Pallikaranai marshland at dawn and flying back north to their roost at dusk. The monsoon frogs which come alive after the rains—cricket frogs, painted frogs, common Indian toads—and sing into my consciousness early in the morning as I sit to meditate. Then the sweat bees and paper wasps which nest in my balcony, in the soil and on the plants, whose activities have anchored my attention through the years.

My gratitude to other nature-educator friends doing phenomenal work, connecting children and people to the local living world and healing the earth's connective tissue—Vena, Sudha, Elizabeth, Surya, Nisha, Ramnath, Rachita, Faiza, Gerry, Chandini, Sharan, Sujatha, Gowri, Ranjani, Prem and Aneesa among others. To thousands of children I've worked with and whose energy and keenness for all of life has guided me, especially among them Amiga, Tvisha, Raoul and Mithul. And to the Greater Chennai Corporation and to Deputy Commissioner Sharanya Ari, who has invited us to implement nature-education pedagogy in Chennai's corporation schools.

To my activist friends alongside whom I have fought to save many wild places in and around Chennai—Prashanth, Vishvaja, Saravanan, Pooja, Divya and Durga, among others.

My advocate friends Yogesh and Poonkuzhali who continue to legally defend nature on a daily basis and have helped me as well.

My thanks to the waterbirds of Vedanthangal Bird Sanctuary—painted storks, openbill storks, black-headed ibises and egrets, among others—which come here in tens of thousands to nest every monsoon. The flying foxes which roost at the Anna Tower Park, the IIT campus, the Theosophical Society and the Independence Day Park. The flamingoes of Pulicat. The white-bellied sea eagle pair of Ennore, which have been resisting industrial eviction for years.

Gratitude to my writer friends whose books, essays and conversations have variously been formative—Neha, Bijal, Siddharth, Cara, Zai, Arati, Bahar, and Shridhar among others.

Then to my Instagram follower-friends and their consistent years-long engagement with and encouragement for my writing on that medium. Gratitude to Sivapriya, whose confidence in my writing and whose kindness as an editor made this book possible.

And to Sarah of Ithaka, whose keenness in the book and efforts to publish work from the Global South has carried *Intertidal* across seas.

My many many thanks to Kanishka Gupta, who has been an exceptional agent and friend.

Thanks to Mugdha, who is among the best parents I have known, and an extraordinary artist. She made this book's cover.

Gratitude to my sister Yazhini who, in her very short life, taught me a great deal about sensitivity and generosity. And to my late uncle Charles Lawrence, who saw me as his own son, in the absence of a caring father. Thanks to the people of my family who have shown up and cared for me—my aunts Esther and Helen, my cousin Vimmi and her husband Santhakumar, my uncle William and his wife Malini, and my cook Krishnaveni.

Profound gratitude to the artisanal fisherfolk of Urur Kuppam for their friendship, their company and their collaborations. They have taught me so much from the intertidal, second only to the ocean and the coast. Then to the Palar and Kosasthalayar rivers which over millennia have crafted this landscape/waterscape whose converging basins I inhabit, and the Chennai coast where they meet the Bay of Bengal.

ABOUT THE AUTHOR

Credit: TEDx Napier Bridge

Yuvan Aves is a writer, naturalist, nature-educator and activist based in Chennai, on the South Indian coast. He writes at the intersection of ecology, education and human/more-than-human consciousness. He is the author of two nature-writing books and three children's books, recipient of the M. Krishnan Memorial Nature Writing Award and the Sanctuary Asia Green Teacher Award among others. He is an active part of resistance and community movements against ecocide and industrial violence in India. He is the founder–managing trustee of Palluyir Trust for Nature Education and Research.